CLIMATE CHANGE AND ITS CAUSES, EFFECTS AND PREDICTION

COUNTRIES AND CLIMATE: POLICIES AND PATHS TO CHANGE

CLIMATE CHANGE AND ITS CAUSES, EFFECTS AND PREDICTION

Additional books in this series can be found on Nova's website under the Series tab.

Additional E-books in this series can be found on Nova's website under the E-books tab.

CLIMATE CHANGE AND ITS CAUSES, EFFECTS AND PREDICTION

COUNTRIES AND CLIMATE: POLICIES AND PATHS TO CHANGE

THOMAS P. PARKER
EDITOR

Nova Science Publishers, Inc.
New York

Copyright © 2011 by Nova Science Publishers, Inc.

All rights reserved. No part of this book may be reproduced, stored in a retrieval system or transmitted in any form or by any means: electronic, electrostatic, magnetic, tape, mechanical photocopying, recording or otherwise without the written permission of the Publisher.

For permission to use material from this book please contact us:
Telephone 631-231-7269; Fax 631-231-8175
Web Site: http://www.novapublishers.com

NOTICE TO THE READER

The Publisher has taken reasonable care in the preparation of this book, but makes no expressed or implied warranty of any kind and assumes no responsibility for any errors or omissions. No liability is assumed for incidental or consequential damages in connection with or arising out of information contained in this book. The Publisher shall not be liable for any special, consequential, or exemplary damages resulting, in whole or in part, from the readers' use of, or reliance upon, this material. Any parts of this book based on government reports are so indicated and copyright is claimed for those parts to the extent applicable to compilations of such works.

Independent verification should be sought for any data, advice or recommendations contained in this book. In addition, no responsibility is assumed by the publisher for any injury and/or damage to persons or property arising from any methods, products, instructions, ideas or otherwise contained in this publication.

This publication is designed to provide accurate and authoritative information with regard to the subject matter covered herein. It is sold with the clear understanding that the Publisher is not engaged in rendering legal or any other professional services. If legal or any other expert assistance is required, the services of a competent person should be sought. FROM A DECLARATION OF PARTICIPANTS JOINTLY ADOPTED BY A COMMITTEE OF THE AMERICAN BAR ASSOCIATION AND A COMMITTEE OF PUBLISHERS.

Additional color graphics may be available in the e-book version of this book.

LIBRARY OF CONGRESS CATALOGING-IN-PUBLICATION DATA

Countries and climate : policies and paths to change / Thomas P. Parker, editor.
 p. cm.
 Includes index.
 ISBN 978-1-61728-923-1 (hardcover)
 1. Climatic changes--Government policy. 2. Climatic changes--International cooperation. 3. Greenhouse gas mitigation. 4. Climatic changes--Government policy--United States. I. Parker, Thomas, 1947-
 QC903.C68 2010
 363.738'740561--dc22
                                       2010026754

Published by Nova Science Publishers, Inc. † New York

CONTENTS

Preface		vii
Chapter 1	An Overview of Greenhouse Gas (GHG) Control Policies in Various Countries *Jane A. Leggett, Richard K. Lattanzio, Carl Ek and Larry Parker*	1
Chapter 2	Climate Change and EU Emissions Trading Scheme (ETS): Looking to 2020 *Larry Parker*	65
Chapter 3	A. U.S.-Centric Chronology of the International Climate Change Negotiations *Jane A. Leggett*	91
Chapter 4	Opening Statement of Chairman Edward J. Markey, The Select Committee on Energy Independence and Global Warming, Hearing on "Preparing for Copenhagen: How Developing Countries are Fighting Climate Change"	111
Chapter 5	Testimony of Carter Roberts, President and CEO, World Wildlife Fund, before the Select Committee on Energy Independence and Global Warming, Hearing on "Preparing for Copenhagen: How Developing Countries are Fighting Climate Change"	113

Chapter 6	Testimony of Barbara A. Finamore, Senior Attorney and China Program Director, Natural Resources Defense Council, President, China-U.S. Energy Efficiency Alliance, before the Select Committee on Energy Independence and Global Warming, Hearing on "Preparing for Copenhagen: How Developing Countries are Fighting Climate Change"	131
Chapter 7	Testimony of Ned Helme, President, Center for Clean Air Policy, before the Select Committee on Energy Independence and Global Warming, Hearing on "Preparing for Copenhagen: How Developing Countries are Fighting Climate Change"	141
Chapter 8	Statement of Lee Lane, Resident Fellow at the American Enterprise Institute, before the Select Committee on Energy Independence and Global Warming, Hearing on "Preparing for Copenhagen: How Developing Countries are Fighting Climate Change"	153
Chapter Sources		175
Index		177

PREFACE

As Congress considers legislation to address climate change, and follows negotiations toward a new international agreement to reduce greenhouse gas (GHG) emissions, the question of the comparability of actions across countries frequently arises. Concerns are raised about what the appropriate sharing of efforts should be among countries, as well as the potential trade implications if countries undertake different levels of GHG reductions and, therefore, incur varying cost impacts on trade-sensitive sectors. All countries examined have in place, or are developing, some enforceable policies that serve to reduce GHG emissions. This book presents an overview of GHG control policies within effect or under consideration in a number of large countries and offers a brief set of intitial observations

Chapter 1- As Congress considers legislation to address climate change, and follows negotiations toward a new international agreement to reduce greenhouse gas (GHG) emissions, the question of the comparability of actions across countries frequently arises. Concerns are raised about what the appropriate sharing of efforts should be among countries, as well as the potential trade implications if countries undertake different levels of GHG reductions and, therefore, incur varying cost impacts on trade-sensitive sectors. This report summarizes the GHG control policies in effect or under consideration in the European Union (EU) and various other large countries, and offers a brief set of initial observations. It gives particular emphasis to how particular trade- sensitive sectors may be treated in the context of each national program.

Chapter 2- The European Union's (EU) Emissions Trading Scheme (ETS) is a cornerstone of the EU's efforts to meet its obligation under the Kyoto Protocol. It covers more than 10,000 energy intensive facilities across the 27

EU Member countries; covered entities emit about 45% of the EU's carbon dioxide emissions. A "Phase 1" trading period began January 1, 2005. A second, Phase 2, trading period began in 2008, covering the period of the Kyoto Protocol. A Phase 3 will begin in 2013 designed to reduce emissions by 21% from 2005 levels.

Several positive results from the Phase 1 "learning by doing" exercise assisted the ETS in making the Phase 2 process run more smoothly, including: (1) greatly improving emissions data, (2) encouraging development of the Kyoto Protocol's project-based mechanisms—Clean Development Mechanism (CDM) and Joint Implementation (JI), and (3) influencing corporate behavior to begin pricing in the value of allowances in decision-making, particularly in the electric utility sector.

Chapter 3- Under the 2007 "Bali Action Plan," countries around the globe sought to reach a "Copenhagen agreement" in December 2009 on effective, feasible, and fair actions beyond 2012 to address risks of climate change driven by human-related emissions of greenhouse gases (GHG). The Copenhagen conference was beset by strong differences among countries, however, and (beyond technical decisions) achieved only mandates to continue negotiating toward the next Conference of the Parties (COP) to be held in Mexico City in December 2010. The COP also "took note of" (not adopting) a "Copenhagen Accord," agreed among the United States and additional countries (notably including China), which reflects compromises on some key actions.

Chapter 4- Over the last two years, the Select Committee has examined closely how the U.S. can fight climate change and improve our energy security. But we are not in this fight alone, and the progress that our country can make is deeply dependent on the progress that developing countries are making. That is the focus of today's hearing: to take an assessment based on the facts that exist in 2009 – not as they existed five or ten years ago – of steps taken by the key developing countries to address global warming.

This inquiry is important because Americans rightly want to know that they are not the only ones altering their policies to combat global warming. This inquiry is also important because many Members have rightly expressed concern about maintaining the competitiveness of critical industry sectors, and they want to know that other countries are joining the fight and requiring their industries to move away from business as usual.

Chapter 5- Chairman Markey, Ranking Member Sensenbrenner, Members of the Committee: On behalf of World Wildlife Fund (WWF), I am pleased to present testimony to this committee. First, let me commend the Chairman and the Select Committee for its important work in bringing much-needed attention within the Congress to so many aspects of climate change. The many hearings held by this committee puts the Congress and the United States as a whole in a much better position to support the domestic legislation and international agreements necessary to respond to this global crisis. So thank you for your important leadership.

Chapter 6- Chairman Markey, Ranking Member Sensenbrenner, and distinguished Members of the Committee, it is my pleasure to be here with you today to discuss China's national greenhouse gas mitigation efforts and achievements. I applaud the committee for calling a hearing on the vitally important topic of how developing countries, including China, are already taking action to fight climate change.

Beginning with its Eleventh Five-Year Plan, which covers 2006 to 2010, China has recognized that it must reduce its rapid growth in energy demand and greenhouse gas emissions and accordingly has embarked on what President Obama in his speech to Congress last week called "the largest effort in history to make their economy energy efficient." It is important that the United States understand what measures China is taking to reduce its greenhouse gas emissions as well as how the United States can strengthen its engagement with China on climate change, because China and the United States together are the two countries that can have the greatest impact on mitigating climate change.[1] The Chinese viewed Secretary of State Clinton's recent visit to Beijing and her message of cooperation extremely favorably and are eager to find areas for mutual cooperation.

Chapter 7- Mr. Chairman, Ranking Member Sensenbrenner and Members of the Committee: I would like to thank you for the opportunity to testify before you today. My name is Ned Helme and I am the President of the Center for Clean Air Policy (CCAP), a Washington, DC and Brussels-based environmental think tank with on the ground programs in New York, San Francisco, Mexico City, Beijing, Jakarta and many other places.

Since 1985, CCAP has been a recognized world leader in climate and air quality policy and is the only independent, non-profit think-tank working exclusively on those issues at the local, national and international levels. CCAP helps policymakers around the world to develop, promote and implement innovative, market-based solutions to major climate, air quality and energy problems that balance both environmental and economic interests.

Chapter 8- Mr. Chairman, Mr. Sensenbrenner, other members of the Committee, thank you for the opportunity to appear before you today. I am Lee Lane, a Resident Fellow at the American Enterprise Institute. AEI is a non-partisan, non-profit organization conducting research and education on public policy issues. AEI does not adopt organizational positions on the issues that it studies, and the views that I express here are mine, not those of AEI.

Rising amounts of greenhouse gases (GHGs) in the atmosphere pose worrisome challenges. While many uncertainties persist, I believe that the potential risks from climate change could be large. At the same time, a thicket of intractable problems blocks quick or easy solutions. Progress on climate policy will require us to wrestle with these problems over many, many decades. My statement suggests some ways in which the US might make progress on this task. It makes three main points.

In: Countries and Climate Policies and Paths... ISBN: 978-1-61728-923-1
Editors: Thomas P. Parker © 2011 Nova Science Publishers, Inc.

Chapter 1

AN OVERVIEW OF GREENHOUSE GAS (GHG) CONTROL POLICIES IN VARIOUS COUNTRIES

Jane A. Leggett, Richard K. Lattanzio, Carl Ek and Larry Parker

SUMMARY

As Congress considers legislation to address climate change, and follows negotiations toward a new international agreement to reduce greenhouse gas (GHG) emissions, the question of the comparability of actions across countries frequently arises. Concerns are raised about what the appropriate sharing of efforts should be among countries, as well as the potential trade implications if countries undertake different levels of GHG reductions and, therefore, incur varying cost impacts on trade-sensitive sectors. This report summarizes the GHG control policies in effect or under consideration in the European Union (EU) and various other large countries, and offers a brief set of initial observations. It gives particular emphasis to how particular trade- sensitive sectors may be treated in the context of each national program.

All countries examined have in place, or are developing, some enforceable policies that serve to reduce GHG emissions. Most are at some stage of making their programs more stringent. The wealthiest countries have all taken on GHG limitation or reduction targets under the Kyoto Protocol. Some of the emerging economies have voluntarily stated GHG targets, though none have yet accepted legally binding obligations in an international agreement. The forms of targets, and their stringencies, vary widely across countries.

The scope of specific GHGs and economic sectors covered by national (or sub-national) reduction measures is generally, but not completely, similar. All have policies that affect carbon dioxide emissions; most have some measures that cover the additional five gases covered under the Kyoto Protocol (methane, nitrous oxide, sulfur hexafluoride, perfluorocarbons, and hydrofluorocarbons).

The programs and measures used vary across countries. Even when some measures have similar names (e.g., voluntary programs and voluntary action plans), the measures may differ in important ways that may influence their effectiveness and impacts on trade competiveness. Within sectors of a country, emission rates and control requirements may vary widely. A country may have some facilities with emission rates (or energy intensities) comparable to the best globally, even if the country's sectoral average as a whole has, for example, a significantly higher energy intensity than the global average.

This report presents an overview of GHG control policies within individual countries. It does not present a rigorous assessment of the comparability of GHG control policies across countries or within specific sectors. The criteria for assessing comparability internationally are not widely agreed, and could encompass a range of considerations, not all quantitatively measurable.

This report summarizes the greenhouse gas (GHG) control policies in effect or under consideration in a number of large countries, and offers a brief set of initial observations. This overview allows preliminary comparison across countries. Because of congressional interest in the comparability of countries' actions, and in the potential trade ramifications of differential policies, these country fact sheets give emphasis to how particular trade-sensitive sectors may be treated in the context of each national program. Where specific industries are not listed in a country's fact sheet, no further information was found.

The European Union's policies are presented first, followed by any additional rules or policies under consideration in several of the largest EU Member States (i.e., France, Germany, the United Kingdom). A number of additional large-emitting countries follow in alphabetical order. Finally, the Appendix provides a comparison of early 2009 vehicle efficiency standards across countries, which may be a useful reference for a sector that emits a large portion of global GHG emissions.

SYNTHESIS OBSERVATIONS

- All countries examined have in place, or are developing, some enforceable policies that serve to reduce greenhouse gas (GHG) emissions. Most are at some stage of making their programs more stringent.
- The scope of specific GHGs and economic sectors covered by national (or sub- national) reduction measures is generally, but not completely, similar. All have policies that affect carbon dioxide (CO_2) emissions; most have some measures that cover the additional five gases covered under the Kyoto Protocol, including methane (CH_4), nitrous oxide (N_2O), sulfur hexafluoride (SF_6), perfluorocarbons (PFC), and hydrofluorocarbons (HFC).
- The programs and measures used vary across countries. Even when some measures have similar names (e.g., voluntary programs and voluntary action plans), the measures may differ in important ways that may influence their effectiveness and impacts on trade competiveness. For example, many countries support "voluntary programs" or "voluntary action plans." Some of these voluntary efforts may provide technical assistance with few requirements from participants; other programs may include formal emission reduction targets, reporting, and governmental pressure to achieve targets.
- Within economic sectors of a country, emission rates and control requirements may vary widely. A country may have some facilities with emission rates (or energy intensities) comparable to the best globally, even if the country's sector as a whole has, for example, an energy intensity significantly higher than the global average for that sector. Such discrepancies often occur in emerging economies wherein an older, less-efficient industrial sector is being replaced by new infrastructure.
- Most of the programs include provisions to assist or exempt trade-sensitive sectors, but the definition of what is trade-sensitive, and the approaches to assisting or protecting the sectors, vary widely. "Trade-sensitivity" is a continuing phenomenon. Companies become more or

less competitive on an international market according to a host of factors, including productivity, market demand, resource costs, labor costs, exchange rates, and the like. The addition of a carbon control regime to this competitive dynamic has raised concerns that, in the absence of similar policies among competing nations, trade-exposed industries that must control their emissions, or face increased costs passed- through by suppliers, may be less competitive and may lose global market share to competitors in countries lacking comparable carbon policies.[1] These concerns have led many countries to consider specific provisions for exposed sectors.

- Assessing the comparability of GHG control policies across countries and in specific sectors could be difficult, and the results could be subject to debate. How well alternative policy directions and methods could stand up under possible challenges against border adjustments under the World Trade Organization (WTO) may merit further investigation. However, consideration of specific methods to assess comparability, and their implications, is beyond the scope of this report.

EUROPEAN UNION[2]

1. Overall GHG Emission Target, if any, and Timing

Under the Kyoto Protocol, the European Union (EU) agreed to reduce GHG emissions of its 15 Member states in 1997 (EU-15) in aggregate by 8% below 1990 levels during the first commitment period of 2008-2012. (There is no collective target for the EU-27, the current 27 Member states of the EU.) In 2007 and 2008, EU-15 GHG emissions were approximately 5% and 6%, respectively, below 1990 levels. In November 2009, The European Commission projected that the EU-15 will surpass its obligation to reduce GHG emissions under the Kyoto Protocol.[3] The EU-15 will have reduced their domestic GHG emissions to about 7% below 1990 levels during 2008-2012. Plans by EU-15 Member states to acquire international credits through the Kyoto Protocol's three market-based mechanisms would provide another 2.2% GHG reduction, while acquisitions by operators in the EU Emission Trading Systems may provide an additional 1.4% GHG reduction, and enhancement of carbon removals by sinks may offer another 1.0%. With additional policies and measures, the Commission projects that the EU-15 may be around 13% below 1990 levels in 2008-2015.

For the post-Kyoto period (beyond 2012), the European Council adopted on April 23, 2009, the "20-20-20" Policy—a climate and energy package to require by 2020:

- a 20% reduction in GHG emissions from 1990 levels,
- a 20% share of renewable energy in the European Union's final consumption figures (including a 10% share in each Member State's transport sector), and
- a 20% reduction in energy consumption.[4]

The legislation also committed to scale up the GHG emission reduction target to 30% if other developed countries make comparable efforts under a new international agreement. The purpose is to limit the global temperature rise to no more than 2°Celsius above preindustrial levels.

2. Principal Policy Instrument(s)

a. Expansion of current European Union Emissions Trading System (EU ETS).[5]
b. Effort-sharing relationships among Member States to reduce emissions in sectors not covered by the EU ETS. It will be left to Member States to define and implement policies in such sectors, although a number of EU-wide measures in areas such as efficiency standards, passenger car emission standards, and a landfill directive for waste disposal will contribute. The European Community infringement procedures and mechanisms for corrective action under the effort-sharing decision are to be put in place to monitor progress.[6]
c. Regulations stipulating mandatory national targets for the overall shares of energy from renewable sources in gross final consumption of energy, taking into account differing starting points for each Member.[7] It will be left to Member States to determine renewable share allocation among sectors.

At the national level, several EU Member states also impose carbon emission fees to some degree. Carbon fees exist in Denmark, Finland, and Sweden. French President Sarkozy had announced carbon taxes to begin on

January 1, 2010, on French households and motor fuels, though their introduction has been delayed by an adverse constitutional ruling. Spain and Ireland reportedly have also signaled that they may consider domestic carbon fees in addition to EU and other national policies.[8] In addition, on October 5, 2009, an EU Taxation Commissioner revealed that in early 2010 the European Commission plans to propose an expansion of existing energy taxation in order to charge CO_2 emission fees as well.[9] The new carbon tax would cover sectors not under the EU ETS (see below), such as agriculture, households, and transport. The proposal explicitly is intended to help the EU achieve compliance with its law to reduce GHG emissions to 20% below 1990 levels by 2020. All taxation proposals, to pass into law, require unanimous agreement of the 27 EU Member states, which may be difficult to achieve, and the assent of the European Parliament.

3. Covered Gases and Sectors

The only greenhouse gas covered under the original 2003 EU ETS was CO_2. The expanded EU ETS to take effect in 2013 will add N_2O emissions from nitric, adipic, and glyoxalic acid production, and PFC emissions from the aluminum sector. Gases not stipulated in the EU ETS, but defined as "greenhouse gases" in Annex II of DIRECTIVE 2003/87/EC include CH_4, HFC, and SF_6. These gases will be controlled under guidelines for sectors not covered by the EU ETS.

Sectors originally covered in the 2003 EU ETS were: power and combustion installations (exceeding 20 megawatts (MW)); petroleum refineries; coke ovens; metal ore production installations; iron and steel production installations (exceeding 2.5 tons of product per hour); factories for cement (exceeding 50 tons per day), glass (exceeding 20 tons per day); ceramics including tiles, bricks, stoneware, porcelain (exceeding 75 tons per day); and production of pulp, paper and board (exceeding 20 tons per day). The expanded EU ETS will increase the scope of covered sectors beginning in 2013 to include primary and secondary aluminum production facilities; ferrous, ferro-alloy, and non-ferrous metal production facilities; mineral wool and gypsum plants; ammonia, petro-chemical and chemical plants including carbon black organics, nitric acid, adipic acid, glyoxal, organic chemicals (exceeding 100 tons per day), hydrogen (exceeding 25 tons per day), soda ash, and sodium bicarbonate. Additionally, certain categories of aviation will be incorporated into the ETS involving commercial flights departing or arriving in a territory of a Member State.[10] In the EU ETS, Member states decide a National Allocation Plan (NAP), subject to review by the EU, to give emission allowances to individual plants. In the first pilot trading period, some Member states allocated more emission allowances than needed to companies, so that revisions to the scheme in Phase III, beginning in 2013, have been adopted to avoid over-allocation, including increasing rates of auctioning allowances.

Sectors not covered by the EU ETS but covered by adopted legislation include transport, housing, agriculture, and waste (see the following discussion).

4. Allocation of GHG Reductions to Various Sectors

The European Union's programs call for a 21% reduction in EU ETS sector emissions compared to 2005 and a 10% reduction in non-EU ETS sector emissions compared to 2005. This is expected to achieve an overall reduction of 14% compared with 2005, which is equivalent to a reduction of 20% compared with 1990. The EU ETS covers electricity generation and the main energy-intensive industries—power stations, refineries, iron and steel, cement and lime, paper, food and drink, glass, ceramics, engineering, and vehicles. Initially, countries allocate allowances to covered sectors, but limited auctioning of permits is planned for the future (e.g., maximum 10% of allowances are auctioned in Phase II).

Phase III ETS: Emissions from sectors covered in the EU ETS will be cut 21% from 2005 levels by 2020. A single EU-wide cap on emissions will be set for EU ETS covered sectors. Allowances will be allocated on the basis of rules harmonized across Member states. The tentative annual cap figure will begin at 1,974 million tons CO_2 in 2013 and decrease annually. The total number of allowances (one allowance equals permission to emit one ton) in 2013 will begin at the average total quantity issued for the 2008-2012 period and will decrease annually at a rate of 1.74%. Free allocation of emission allowances will be progressively replaced by auctioning allowances by 2020. Auctioning will begin in 2013 at 20% and gradually rise to 70% in 2020 and to 100% in 2027. Power producers must acquire all allowances at auction in order to prevent windfall profits (following experience under the pilot trading period). Member States that are highly dependent on fossil fuels and/or States insufficiently connected to the grid (these include Bulgaria, Cyprus, Czech Republic, Estonia, Hungary, Latvia, Lithuania, Malta, Poland and Romania) are allowed to apply for a derogation procedure of reduced auctioning rates for power production of 30% in 2013, gradually rising to 100% in 2020, as long as producers invest in clean technologies to the market value of the permits. Furthermore, less affluent states (the 10 above plus Greece and Portugal) will receive an increased amount of emission permits to auction amounting to 12% more than their actual share to assist in revenue generation. Each Member state will be allowed to determine use of revenue with a suggested investment of 50% toward clean technologies and pollution abatement.

Non-ETS: Sectors not covered by the EU ETS are transport, housing, agriculture and waste. The 2009 Directive proposes to cut emission in these sectors by 10% EU-wide from 2005 levels by 2020. Targets will be mandated according to each Member states' relative wealth (based on GDP per capita and economic growth prospects) with figures ranging from -20% to +20%. Targets are binding on Member states and are enforceable through the usual EU infringement procedure.[11] If a country exceeds its annual objective, it must implement corrective measures, and will be penalized via a deduction from the following year's CO_2 allowance. Several flexibility measures are available including the possibility of trading emission cuts across countries; carrying forward ("banking") extra emission reductions; and using a limited amount of credit from developing countries (through an offsets mechanism similar to the Kyoto Protocol's Clean Development Mechanism).

The transportation sector has legally binding standards for CO_2 emissions from new passenger cars to apply as of 2012 in order the meet the 20% emission reduction by 2020.[12] Reductions are required to achieve 120 grams carbon dioxide per kilometer (CO_2/km) for 65% of fleet in 2012, 75% in 2013, 80% in 2014 and 100% starting in 2015. A target of 95 grams CO_2/km is set for 2020. Enforcement is set through financial penalties against the car manufacturers depending on how far their fleet exceeds the targets.[13]

A renewable energy mandate sets mandatory national targets for each Member state in accordance with each country's different starting points. The purpose of mandates is to provide certainties for investment. Each country will report to the European Council by June 2010 regarding how each Member has allocated the renewable target among transport, electricity, heating and cooling sectors. A 10% target for renewable energy in the transportation sector is set at the same level for all countries.

5. Any Regulations or Exemptions Specific to Trade-sensitive Sectors

The climate and energy package in the 2009 Directive provides that the risk of "carbon leakage"[14] may be reduced by allotting free carbon allowances to businesses exposed to "significant risk of carbon leakage" (SRCL) by the cost of compliance with the EU ETS. (The European Commission must adopt a list of sectors deemed exposed to a significant risk of carbon leakage no later than December 31, 2009. A draft list was proposed in September 2009, discussed below.) However, any free allowances will not be decided until 2011. The list may be revised before 2014, based on reanalysis of trade figures, and identification of countries that make firm commitments to reduce their GHG emissions.

If international negotiations on climate change in Copenhagen do not lead to a comprehensive international agreement, several criteria permit an EU ETS-covered industrial sector to allege SRCL:

- if the industry can demonstrate that purchasing permits increases its costs (more than 5% of gross value added) and faces international competition (non-EU trade intensity above 10%), or
- if the industry can demonstrate that purchasing permits significantly increases its costs (more than 30% of gross value added), or
- if the industry faces international competition (non-EU trade intensity above 3 0%), then it can qualify for the free allocation of allowances.

Free allocation of permits typically will not be at 100% of needs for SRCL facilities, however. Free allowances will be adjusted according to Community-wide ex-ante benchmarks so as to ensure incentives for GHG reduction. The benchmarks will be set at the average performance of the 10% most GHG emissions-efficient installations in a sector in 2007-2008. Only the most efficient businesses in a sector, therefore, have a chance to receive all of their allowances free. If a business emits more than this benchmark allocation, it will need to acquire allowances up to its actual emissions.

As of September 2009, EU analysis assessed the industries and productions potentially exposed to carbon leakage risks. Assuming that 100% of allowances were auctioned (which will not occur initially), the analysis concluded that 146 sectors (out of 258) and five additional product categories meet the EU's criteria for being exposed to SRCL.[15] Outside of these sectors, 13 subsectors and products may be exposed to risk: food processing industries;

industrial gases; nonmetallic mineral products; glass fibers (filament glass fibers); and, colors and similar preparations for ceramics/glass etc.[16] The EU analysis estimates that the listed sectors now constitute about 75% of GHG emissions covered by the EU ETS.

An alternative approach to issues of competitiveness in trade sensitive sectors put forward by the European Commission is the integration of importers into the EU ETS. Under an integrated emission trading regime, foreign producers would purchase emission certificates for their imports according to the emissions produced. In a speech in London on January 21, 2008, the President of the European Commission, Jose Manuel Barroso, said: "I think we should also be ready to ... require importers to obtain allowances alongside European competitors, as long as such a system is compatible with WTO requirements." Beyond these measures, French President Nicolas Sarkozy, with possible interest from German Chancellor Angela Merkel, has indicated interest in potentially charging carbon levies against imports from countries that do not meet stringent environmental standards. (See fact sheet on France. See also the Appendix, comparing EU efficiency standards for motor vehicles with those of other countries.)

On December 22, 2009, E.U. environment ministers—including those from then-E.U. President Sweden and incoming-E.U. President Spain—were stated as saying that the 27-nation bloc would consider imposing carbon tariffs and other border sanctions in the wake of perceived failures at the Copenhagen climate conference. No details were stated, but Teresa Ribera, Spain's Secretary of State for Climate Change, reportedly said that Spain plans to convene special meetings for the E.U. environmental ministers in the upcoming months to discuss the "strategic line" the European Union should take in promoting an international environmental agenda.

FRANCE[17]

(Policies and statements if substantially different from the European Commission)

1. Overall GHG Emission Target, if any, and Timing

Under the Kyoto Protocol, France's share of the EU target is not to exceed the 1990 level during the period 2008-2012.

France has a stated long-term national GHG emissions target of 75% below the 1990 level by 2050. A law is planned to reduce energy consumption of existing buildings by 38% by 2020.

2. Principal Policy Instrument(s) (See "EU ETS.")

Beyond instruments of the European Union, policy considerations have ranged from a freeze on the building of new highways and airports, to a vast plan to shift freight traffic from road to rail, to a commitment to slash pesticide use by half within 10 years by Europe's biggest farm producer. Tramway and TGV high-speed train networks are to be extended, and drivers encouraged to buy cleaner cars through bonuses and penalties. In October 2007, French President Nicolas Sarkozy called for a plan to institute a national "carbon tax" on global-warming pollutants. The Sarkozyproposed carbon tax was rejected by France's Constitutional Council in December 2009; however, the party said the measure would be redrafted for passage in 2010 (see below).

3. Covered Gases and Sectors (See "European Union.")

The administration's proposed carbon tax would apply to households and motor fuels but not to large businesses and power generators, as they are not covered by the EU ETS.

4. Allocation of GHG Reductions to Various Sectors

About half of French industry's GHG emissions are covered by the EU ETS, including large emission sources in the power generation, iron, steel, glass, cement, pottery and brick sectors.

In September 2009, President Nicolas Sarkozy stated that the proposed carbon tax would begin in January 2010. Because Sarkozy's party holds a majority in its parliament, expectations are that the new carbon levy will be enacted into law. Initially set at 17 Euros (US$25)[18] per ton of emitted CO_2, the tax on the use of oil, natural gas and coal would nudge up the cost of a liter of gasoline by US$0.06 (US$0.23 a gallon). It would apply to households as well as enterprises, but not to the heavy industries and power companies in France that are covered by the EU's emissions trading scheme (see the EU ETS under "European Union"). Revenues from the new tax would be returned to taxpayers through cuts in income tax and other taxes. France's Le Monde newspaper says the tax will cover 70% of the country's carbon emissions (e.g., from vehicles) and bring in about 4.3 billion Euros (US$6.4 billion) of revenue annually. Sweden, Denmark, Finland, Norway and Switzerland already impose similar taxes, although Sweden's is levied at a much higher emission fee (108 Euros/ton of CO_2, or US$161/ton).

On December 28, 2009, France's Constitutional Council rejected the carbon tax because the bill reportedly contained too many exceptions for polluters, broke with past practices, and produced an unfair tax burden on individual consumers. The French Council was stated as saying that the tax was flawed because it would have raised the cost of vehicle and home heating fuel without commensurate increases on other sources of emissions. The Sarkozy administration promised an amended bill back to council ministers by January 20, 2010. To address objections, a new bill would need to subject corporate industry to the tax, a requirement opposed by French companies already concerned with decreasing competitiveness. France continues to back efforts to introduce an E.U.-wide carbon tax and a border tax at E.U. frontiers as ways to allay industry concerns (see below).

5. Any Regulations or Exemptions Specific to Trade-sensitive Sectors

French President Nicolas Sarkozy has promoted a European levy on carbon-intensive imports from countries outside the Kyoto Protocol. The United States could be subject to such proposed fees should it not adopt legally enforceable GHG controls domestically. The Economist has said, "That leads some to suspect that his ultimate objective is to create a pretext for protectionism."[19,20]

In addition, President Sarkozy, along with German Chancellor Angela Merkel, has called for the United Nations to support "appropriate adjustment measures" to be levied against countries that do not join or implement an international agreement being negotiated for agreement in Copenhagen in December 2009.[21]

Motor Vehicles: A law is planned to cut GHG emissions from transport by 20% by 2020; it would include a goal of 7% bio-fuels by 2010 and EU emissions limit for new cars—130g/km— to be phased in from 2012.

GERMANY[22]

(Policies and statements if substantially different from the European Commission)

1. Overall GHG Emission Target, if any, and Timing

Under the Kyoto Protocol, Germany's share of the EU target is to reduce GHG emissions to 21% below 1990 levels during the period 2008-2012. (Germany was able to take on such a deep target because of its reunification with East Germany, taking on East Germany's high emissions baseline and reducing emissions by closing and improving many inefficient installations.)

The German government approved a new package of climate change measures in June 2008 that are a legal transposition of the EU's Integrated Climate Change and Energy Programme.[23] The German measures aim at a CO_2 emission reduction of 40% by 2020 compared to 1990 levels. The legislative package focuses on the transport and construction sectors.

2. Principal Policy Instrument(s) (See "EU ETS.")

The Integrated Climate Change and Energy Programme: In 2007, the German government, working from the general guidelines of European policy decisions, implemented a concrete program of measures at the national level. Through 29 measures, the program addresses a wide range of matters, including combined heat and power generation, the expansion of renewable energies in the power sector, carbon capture and sequestration (CCS) technologies, smart metering, clean power station technologies, the introduction of modern energy management systems, support programs for climate protection and energy efficiency (apart from buildings), energy efficient products, provisions on the feed-in of biogas to natural gas grids, an energy savings ordinance, a modernization program to reduce CO_2 emissions from buildings, energy efficient modernization of social infrastructure, the Renewable Energies Heat Act program for the energy efficient modernization of federal buildings, a carbon dioxide strategy for passenger cars, the expansion of the bio-fuels market, reform of vehicle tax on the basis of carbon dioxide, energy labeling of passenger cars, the reduction of emissions of fluorinated greenhouse gases, procurement of energy efficient products and services, energy research and innovation, increased electric mobility, international projects on climate protection and energy efficiency, reporting on energy and climate policy by German embassies and consulates, and a transatlantic climate and technology initiative. In June 2008, the program was enacted with a package of measures to double electricity generated by combined heat and power technology (CHP) to 25%. The share of renewable electricity will also be increased to 20%, especially through subsidizing off-shore wind farm development. At the same time the package has set a target of producing half of Germany's electricity from renewable energy sources or super-efficient plants by 2020. The package aims for an 11% reduction in electricity consumption by 2020.

Loans for energy efficiency and CO_2 reduction measures in the domestic sector have been available as an economic recovery measure.

3. Covered Gases and Sectors (See "European Union.")

4. Allocation of GHG Reductions to Various Sectors (See "European Union.")

5. Any Regulations or Exemptions Specific to Trade-sensitive Sectors

Germany wants to give companies in globally traded sectors bigger EU allowance quotas in the EU ETS to soften the cost impact of Europe's climate change policy.

Germany has been a vocal opponent of auctioning emissions allowances, although the EU has decided to move forward with limited auctioning. As examples of Germany's past stance, in January 2008, Environment Minister

Sigmar Gabriel critiqued the European Commission's plan to commence auctioning emissions permits that are currently distributed for free, stating that "The European Union cannot ignore the question of how to preserve the international competitiveness of industries that consume lots of energy," such as cement, steel and chemicals, all key sectors of the Germany economy.[24] Sectors "which have reached their average for reductions of carbon dioxide emissions must be able to obtain free emission rights to be able to remain in Europe," claiming that many European industries could be forced to relocate elsewhere in order to maintain competitive prices in international markets. German Economy Minister Michael Glos has also criticized the plan to auction emission rights.[25] Gabriel also condemned the weakness of the commission's project in terms of developing renewable energies, which he said threatened national support for such energies. Gabriel nonetheless reiterated German opposition to EU plans to reduce new car emissions to 120 grams of CO_2/km by 2012 without distinguishing by the class of vehicle (German car makers produce many powerful automobiles which emit high levels of CO_2).

UNITED KINGDOM[26]

(Policies and statements if substantially different from the European Commission)

1. Overall GHG Emission Target, if any, and Timing

Under the Kyoto Protocol, the United Kingdom's (UK) share of the EU target is to reduce GHG emissions to 12.5% below 1990 levels during the period 2008-2012.

Climate Change Act of 2008 introduced a legally binding long-term target to cut emissions by at least 80% by 2050 and at least 34% by 2020 compared to 1990 levels.[27] Major provisions of the act include the setting of legally binding targets, the establishment of a carbon budgeting system, and the creation of a Committee on Climate Change. The carbon budgeting system establishes caps on GHG emissions over five-year periods, with three budget periods being set at a time, charting progress to 2050. The act also requires that the government amend the act to include emissions from shipping and aviation by December 31, 2012. The act states that a reduction of power sector emissions by 40% should be achievable by 2020.

Goal to reduce CO_2 emissions from new houses to zero by 2016.

2. Principal Policy Instrument(s) (See "EU ETS.")

The Carbon Budgeting System is outlined in the 2008 Climate Change Act. In it, the Secretary of State is authorized to set an amount for the net UK carbon account (the "carbon budget") for successive periods of five years each ("budgetary periods"), beginning with the period 2008-2012.

The Carbon Reduction Commitment (CRC)[28] applies to non-energy intensive sectors not covered by the EU ETS. It will apply a mandatory emissions cap and trading program to cut carbon emissions from large commercial and public sector organizations (including supermarkets, hotel chains, government departments, large local authority buildings using more than 6,000 megawatt hours (MWh) of electricity through mandatory half hourly meters) by 1.1 million tons of carbon per year by 2020. Allowances in the CRC system would be sold by auction. The revenue raised from the sale of Carbon Reduction Commitment allowances are to be recycled back into the scheme through bonuses and penalties meant to stimulate organizations to reduce their levels of emissions. Any bonus or penalty administered to an organization are to be based on their ranked position on performance in three metrics (gross emissions, growth, and early compliance actions).[29]

The Carbon Emissions Reduction Target (CERT) came into effect on April 1, 2008, and will run until 2011 as an obligation on energy suppliers to achieve targets for promoting reductions in carbon emissions in the household sector. As reported by the Energy Savings Trust, an independent UK-based non-governmental organization, "it was originally estimated that CERT would stimulate approximately £2.8 billion (US$4.7 billion)[30] of investment by energy suppliers in carbon reduction measures. In September 2008, the Government announced that the level of funding available from the energy suppliers would be increased by £560 million" (US$893 million). The investment would increase the program's lifetime carbon savings to 185 million tons (Mt) CO_2 (31 Mt CO_2 more than under the original CERT target of 154 Mt CO_2).[31]

The Renewable Energy Strategy: The Department of Energy and Climate Change (DECC) details how the UK plans to hit its target of getting 15% of energy (electricity, heat and transport) from renewable sources by 2020. In order to achieve the target, 30% of electricity must come from renewable energy sources, including nuclear power (a five-fold increase from today's rate of ~5%), 12% of heat must be generated by renewables, and 10% of transport energy must be from renewables. The main instrument to achieve these targets for renewable (and nuclear) electricity generation are "Non-Fossil Fuel Obligations" (NFFO), begun in 1989, now Renewables Obligations," requiring operators of the distribution grid to purchase quotas of renewable and nuclear electricity. The prices are subsidized by a Climate Change Levy.[32]

The Climate Change Levy was established in the UK under the Finance Act 2000 (2000 c: 17): a tax on most fuels, including natural gas, electricity (including nuclear) and solid fuels, but not on vehicle or household users, nor renewable energy or cogeneration.[33] Revenues are used to help fund employment insurance, and to fund the Carbon Trust.[34] In addition, energy-intensive businesses qualify for a levy reduced by 80% if they signed voluntary Climate Change Agreements to improve energy efficiency or reduce GHG emissions. Although the Climate Change Levy initially was a fixed rate, the 2006 UK budget tied the rates to account for inflation beginning in 2007.

3. Covered Gases and Sectors (See "European Union.")

4. Allocation of GHG Reductions to Various Sectors

The EU ETS covers electricity generation and the main energy intensive industries—power stations, refineries, iron and steel, cement and lime, paper, food and drink, glass, ceramics, and engineering and vehicles. Overall, these account for around 50% of UK CO_2 emissions. Non- energy intensive, large-scale, commercial and public sectors are covered by the CRC policy (amounting to 25% of the business sector). Household emissions are covered by the CERT policy.[35]

5. Any Regulations or Exemptions Specific to Trade-sensitive Sectors

The UK's "Low Carbon Industrial Strategy" states a vision that the nation "must create the conditions for the UK to be—and be recognised as—the leading location in the world for growing an innovative low carbon business and developing new low carbon products and services."[36] The UK strategy appears oriented toward supporting identified opportunities in "green" businesses and technologies, aiding them through:

- a Low Carbon Investment Fund, (with financing of £405 million—US$674 million);
- a business-led Technology Strategy Board;
- an Energy Technologies Institute (ETI), serving as a private/public partnership to invest in development of low carbon energy technologies;
- R&D tax credits;
- a Carbon Trust to support development and deployment of new and emerging low carbon technologies; and
- a UK "innovation infrastructure," including intellectual property systems and procedures, standards, and a National Measurement System.

AUSTRALIA[37]

1. Overall GHG Emission Target, if any, and Timing

Under the Kyoto Protocol, Australia accepted a target to limit its net GHG emission increase to 8% above 1990 levels. It has also proposed that, under a new international agreement, it would take on a target to reduce its GHG emissions by 5% to 25% below 2000 levels by 2020, with the more stringent commitment conditioned on whether "the world agrees to an ambitious global deal to stabilise levels of CO_2 equivalent in the atmosphere at 450 parts per million (ppm) or lower."[38]

2. Principal Policy Instrument(s)

The Australian government proposed a Carbon Pollution Reduction Scheme (CPRS) to be phased in beginning July 1, 2011. A one-year period would occur from 2011-12, during which carbon emission permits would be sold at a fixed price Aus$10 per ton of carbon (US$9.20);[39] these may not be banked for use in later periods. The full cap-and-trade system would be in effect by 2012, by which time all covered businesses must purchase carbon permits at market prices. The Senate did not pass this proposal on its first or second readings in August and December 2009. Despite addition of several exemptions and aid to selected industries, strong political opposition (including the ousting of the Senate's opposition leader who negotiated the compromise provisions) in the Senate blocked passage of the measure. The Rudd Government has said it will maintain its overall GHG goal of 25% below 2000 levels and will resubmit the proposal to the Parliament again in February 2010. It has also indicated that, if the CPRS does not pass in February, it will lead to "double dissolution" of the Parliament and a snap election, expected to result in increased representation by Rudd's allies.

The Australian program includes a Renewable Energy Target, and investment in carbon capture and storage. Up to 5 percentage points of its offered 25% target for 2020 could be met by purchase of international emission reduction credits using CPRS revenue, though no earlier than 2015. Eligible businesses also may receive government funding for energy efficiency investments, available from a Aus$200 million (US$184 million) portion of a Climate Change Action Fund.

In August, though the Australian Senate did not pass the carbon reduction proposal, it passed the Renewable Energy Target (RET) into law[40] that establishes a system of tradable Renewable Energy Certificates (RECs). It requires that 20% of electricity come from renewable resources by 2020 (projected to require 45 gigawatt hours (GWh)). Currently, about 8% of Australia's electricity is generated with renewables. Among other provisions, the law provides Solar Credits, allowing receipt of a multiple of 2-5 of RECs for qualified installations, that will subsidize the capital costs of small-scale systems, such as household photovoltaic systems. The grants of RECs will depend on the generation of energy, not the installed capacity (which, in some countries, has not stimulated maximizing the use of installed capacity).

3. Covered Gases and Sectors

As proposed, the CPRS would initially cover the six GHG of the Kyoto Protocol, and emissions from stationary energy, transport, industrial processes, waste, forestry, and fugitive emissions from oil and gas production.[41] It is expected to cover 75% of Australia's GHG emissions and about 1000 entities (out of 7.6 million registered businesses in Australia).[42] Agriculture was exempted in order to secure passage of the bill.

4. Allocation of GHG Reductions to Various Sectors

Permits would be available in 2011 at a fixed price of Aus$10 per ton of carbon-equivalent (US$8.60), after which all covered sources must purchase their permits through auction or the market.

5. Any Regulations or Exemptions Specific to Trade-sensitive Sectors

News reports indicate that the Rudd Government's legislative proposal has been modified to gain legislative support for the measure, by exempting agriculture from the system, and increasing financial incentives to electric power generators, coal mines, and food processors.[43] The proposed CPRS includes provisions to assist emissions-intensive, trade-exposed industries (EITE). Eligibility for assistance would be determined by an assessment of all entities conducting a specific activity. First, there would be quantitative and qualitative tests to assess the activity's trade exposure. Second, there would be assessments of greenhouse gas intensity based on the average emissions per million dollars of revenue or emissions per million dollars of value added. The baseline for the emission data would be 2006-2007 to 2007-2008, while the baseline for revenue/value added data would be 2004-2005 to the first half of 2008-2009.

The government allocates free permits using an allocation baseline of emissions per unit of output for each EITE activity. This baseline will provide the basis for eligibility at either the 90% or 60% assistance rates. The proposal[44] would set up two initial rates of assistance: (1) 90% allocation of allowances for activity with emissions intensity of at least 2,000 tons of emissions per million dollars revenue or 6,000 tons of emissions per million dollars of value added; (2) 60% allocation of allowances for activity with emissions intensity between 1,000 tons of emissions per million dollars revenue and 1,999 tons of emissions per million dollars revenue or between 3,000 tons and 5,999 tons of emissions per million dollars of value-added. This assistance per unit of production will be reduced by 1.3% annually.

The proposed CPRS would include a five-year Global Recession Buffer as part of an assistance package to EITE. Industries eligible for 60% assistance would receive a "buffer" of 10% free emission permits; industries eligible for 90% assistance would receive a 5% buffer of free emission permits.

Reviews of the EITE scheme would occur every five years, and would consider a list of identified issues, including whether the assisted firms are making progress toward world's best practice efficiencies, and whether "broadly comparable carbon constraints" are imposed in competing economies. Any changes to the system would require five years' advance notice.

The scope of consideration for assistance includes (1) direct emissions covered, (2) related cost increases for electricity and steam use, and (3) related cost increases for upstream emissions from natural gas and its components (e.g., methane and ethane) used as feedstock. The assistance package would include direct emissions and some indirect emissions.

Two amendment bills to the Renewable Energy (Electricity) Act 2000 were passed on August 20, 2009, and received Royal Assent on September 8, 2009. The Renewable Energy Amendments contain provisions to assist electricity-intensive industries and the coal industry. Under these provisions, one or more emissions-intensive trade-exposed activities may be partially exempted from its REC requirements. If resulting Partial Exemption Certificates are taken into account, it would reduce the charge for falling short of RECs that would otherwise be payable.[45] In this law, the definition of "emissions-intensive trade-exposed activity" would be either defined by further regulations, or by regulations under a Carbon Pollution Reduction Scheme Act 2009 if passed. The methods for calculating the amounts of partial exemptions would be defined by regulations.

BRAZIL[46]

1. Overall GHG Emission Target, if any, and Timing

In November 2009, Dilma Rousseff, chief of staff for Brazilian President Luiz Inácio Lula da Silva, was reported as saying that her country would take a proposal for voluntary GHG emissions reductions of 36%-39% by 2020 to the Copenhagen summit.[47] Brazil's emissions would drop to near 1994 levels if the top end of the pledge is met, representing about a 20% cut from the 2.1 million tons emitted in 2005. The emission cuts would be based largely on reducing deforestation rates, and would depend in large part on obtaining "sufficient" financing. President Lula stated in December 2008 that Brazil would slow its rate of deforestation in the state of Amazonas by 70% by 2017, compared to the average rate from 1996 to 2005. In September 2009, the Brazilian government extended this target to an 80% reduction by 2020.[48] Brazil has set a target by 2010 for zero deforestation in its Atlantic Forest.

On December 28, 2009, President Lula signed into law the 39% reduction in emissions by 2020, meeting the commitment made at the Copenhagen climate conference. The new law, however, is subject to several decrees setting out responsibilities and regulations for the farming, industrial, energy, and environmental sectors, and omits several vetoed provisions, including a reference to "promoting the development of clean energy sources and the gradual phasing out of energy from fossil fuels." President Lula is expected to sign the decrees in January after consulting scientists and other experts.[49]

2. Principal Policy Instrument(s)

In December 2008, Brazilian President Luiz Inácio Lula da Silva signed the National Climate Change Plan (PNMC) into effect.[50] Policy measures include:

- Stimulating energy efficiency through best practice, including the implementation of an energy efficiency policy that targets a savings of 106 terawatt hours per year (TWh/y) by 2030; the substitution of renewable charcoal for coal in manufacturing sectors; the replacement of one million old refrigerators per year for 10 years; the deployment of solar power systems for water heating; and the phasing out of the use of fire for the clearing and cutting of sugarcane.
- Retaining a high renewable energy share in the electricity sector, including the increase of the total electricity supply from cogeneration, mainly from sugarcane bagasse, to 11.4% by 2030; the reduction of non-technical losses in electricity distribution at a rate of 1,000 GWh/y over the next 10 years; the addition of 34,460 MW capacity from new hydropower plants over the next 10 years; the increase in electrical supply share from wind and sugarcane bagasse by 7,000 MW by 2010; and the expansion of the national solar photovoltaic industry and its deployment in systems isolated from the grid.
- Increasing the share of bio-fuels in transport matrix, including the attempt to encourage industry to achieve an annual substitution rate of

11% bio-fuels for fossil sources over the next 10 years; and the institution of a 5% bio-fuel to diesel mandate by 2010;
- Reducing deforestation rates and eliminating forest losses, increasing policing against illegal logging and curtailing financing to illegal ranching.
- Continuing the policy measures of prior renewable energy regulations including the 2004 Program of Incentives for Alternative Electricity Sources (PROFINA), coordinated by the Ministry of Mining and Energy and Centrais Elétricas Brasileiras (Eletrobras). The program contains new strategies for the incorporation of renewable resources in Brazil's energy matrix and strengthens the country's policy on diversification and development. On its inception, PROFINA contracted 144 generation stations to benefit 19 states with a combined capacity of 3,300 MW from wind, biomass, and small hydro sources for a potential GHG reduction of 2.8 Mt CO_2/year.

Many of Brazil's mitigation strategies involve the reduction of deforestation rates in the Amazon. The current administration has expanded protected areas in the Amazon and implemented new environmental policies. More than 62 natural reserves have been established in the Amazon, bringing the total area of the Brazilian Amazon protected by law to 280,000 square kilometers, the fourth-largest percentage of protected area in relation to territory among all countries. In addition to the aforementioned National Climate Change Plan, Brazil has enacted other laws that address deforestation and sustainable development.

- The Public Forest Management Law encourages sustainable development, places a moratorium on soybean plantings and cattle ranching in the Amazon, and authorizes the creation of a plan to reduce the rate of Amazon deforestation by half. Brazil plans to meet this goal by increasing federal patrols of forested areas, replanting 21,000 square miles of forest, and financing sustainable development projects in areas where the local economy depends on logging.
- The Action Plan for the Prevention and Control of Amazon Deforestation intends to improve the monitoring of the deforestation process, from a regional to a local scale; promotes the presence of public authorities in critical zones; confronts the economic speculation problem involved in public lands; plans the appropriate distribution of public lands according to social and ecological needs;

and retains commercial wood exploration while also promoting sustainable forest management.
- The Amazon Fund (a private fund) aims to combat deforestation and to promote sustainable development in the Amazon. In 2008, Norway pledged $1 billion to the fund through 2015, making it the first country to do so, stating that it would donate as much as $130 million in 2009.[51]

The Brazilian government maintains that these efforts have been successful. It has recently been reported that deforestation of the Amazon fell by the largest amount in more than 20 years, dropping 45%, from nearly 5,000 square miles to some 2,700 square miles, in 2008, although there normally is a great deal of year-to-year variability in deforestation rates.[52] A continued emphasis on enforcement coincides with legislation. The enactment of the Prevention of the Use of Illegal Timber in the Building Industry Act, starting January 2009, asks for proof of the legal origin of timber from building companies. As such, the government recovered 1.4 million cubic meters of illegal wood and 700 people were put in prison.[53]

Observers note, however, that other factors contribute to the rate of deforestation beyond governmental policy measures. Brazilian deforestation is strongly correlated to the economic health of the country. Recent reductions are concurrent with the global economic downturn. Falling commodity prices have stalled the expansion of ranching and agriculture into the Amazon. While these trends have seemed favorable for emission reductions, some commentators still point to what they consider continued deforestation practices by commercial and speculative interests, misguided government policies, inappropriate World Bank projects, and commercial exploitation of forest resources. Others see favorable taxation policies, combined with government subsidized agriculture and colonization programs, as a continued encouragement for the destruction of the Amazon. Still others emphasize the inherent difficulty in measuring, reporting and verifying any GHG emission reductions in the Land Use, Land Use Change and Forestry (LULUCF) sector. Finally, most stress the crucial commitment to local law enforcement policies to sustain any regulatory reform that comes out of the federal government.

3. Covered Gases and Sectors

Primarily CO_2 in deforestation and other domestic agendas; however, U.N. Clean Development Mechanism projects in Brazil include CH_4 and N_2O reductions.

4. Allocation of GHG Reductions to Various Sectors

Unlike other developed or developing countries, Brazil holds a unique endowment of natural resources that affects its climate change portfolio in the power generation and transportation fuel sectors. A low contribution of greenhouse gas emissions has been due to both market-driven and governmental decisions to adopt renewable energy sources over the past few decades. The markets for both hydroelectricity and sugarcane products (bagasse for thermal purposes and ethanol for transportation fuel) have expanded 10-fold. During this period there was also an important decrease in wood consumption in the residential and industrial sectors and an increase in charcoal consumption in the industrial sector.

Taken together, however, the sectors of energy, industrial processes, solvents and waste treatment contribute only 25% of total GHG emissions, estimated at approximately 1 billion tons. The rest of Brazilian GHG emissions is tied to the LULUCF sector, and of that total, 90% corresponds to the conversion of forests to other uses, especially agriculture and ranching. For this reason, most of Brazil's mitigation policies have concentrated on the forestry sector.

5. Any Regulations or Exemptions Specific to Trade-sensitive Sectors

Not specified.

CANADA[54]

1. Overall GHG Emission Target, if Any

In April 2007, then-Environment Minister John Baird announced that by 2020, Canada would reduce its GHG emissions by 150 million tons, or 20%, from its 2006 level. Beyond this, the government hopes to achieve a 60%-70% reduction by 2050.[55] The Kyoto emission reduction targets are scored from 1990 (with a few explicit exceptions); some analysts assert that, since Canada's GHG emissions rose 27% between 1990 and 2004, the government would be able to demonstrate far greater progress if it were able to use 2006 as its base year in the Copenhagen Agreement.[56]

2. Principal Policy Instrument(s)

The government's most recent plan for regulating industrial air emissions was announced in March 2008.[57] However, observers note that it remains indefinite. Canada's current Environment Minister, Jim Prentice, is traveling around the country's 10 provinces soliciting ideas on a capand-trade system. There has reportedly been a great deal of pressure on the Minister to develop a plan that will be compatible with whatever may be developed in the United States. For example, the original 2007 Canadian plan called for an "intensity target" rather than a cap. Bilateral discussions over a compatible cap-and-trade system are underway.[58] The effort at cross-border harmonization is likely due to the extensive economic integration between the two countries.

The government aims to complete its policy formulation and present its formal plan before the December 2009 United Nations climate change Conference of the Parties in Copenhagen. Some observers note that the government's ambitions might be delayed or curtailed if a snap election is called; however the prospect of such a vote is believed to be increasingly unlikely.[59]

Recognizing that the transportation sector is responsible for about 27% of GHG emissions, the Canadian government is also set to issue mandatory auto emissions regulations—essentially converting fuel efficiency into CO_2 limits—and likely will seek to make its standards compatible with those set by the U.S. Environmental Protection Agency. The Environment Ministry may also issue modified regulations regarding usage of ethanol. These changes would be facilitated by amendments to the Canadian Environmental Protection Act of 1999, which, among other things, can be used to regulate tailpipe emissions and ethanol blending. Regulations have yet to be published; the ministry likely will attempt to match and harmonize its emissions standards on a continental basis.

The federal government can also use its spending power to control pollution. The government has created a climate change "ecoTrust" fund from which the provinces may draw in order to pay for programs to reduce their own GHG emissions. The last two federal budgets have also included significant funding for carbon capture and storage, including a large-scale demonstration facility. This could be one important aspect of the attempt to reduce emissions arising from some provinces' extensive use of coal as an energy source; it also could be used for oil sands.

3. Covered Gases and Sectors

Although the details are still being negotiated, Canada's regulations will likely cover the six gases included in the Kyoto Protocol. In reducing GHG emissions in Canada, the government will likely also attempt to co-reduce other pollutants such as sulfur dioxide, particulate matter, and mercury. Specific sectors have yet to be determined.

4. Allocation of GHG Reductions to Various Sectors

The government has not yet determined the sectoral allocation of reductions, but it has calculated that 35% of Canada's GHG emissions arise from fossil fuel production, industrial processing and manufacturing; 22% from services, residential, waste and agriculture; 16% from electricity and heat generation; and 27% from transportation. [60]

5. Any Regulations or Exemptions Specific to Trade-sensitive Sectors

Canadian government officials maintain that exemptions—if any—and regulations are yet to come, and that Environment Minister Prentice is still attempting to strike agreements with the various provinces.

CHINA[61]

1. Overall GHG Emission Target, if any, and Timing

The 11th Five-Year Plan set compulsory energy and pollution targets for 2006-2010 that have slowed growth of GHG emissions, and those energy targets appear likely to be reached or surpassed by 2010. However, as China publishes neither its GHG emissions nor the effects of policies on GHG trajectories, validating reports of progress is not possible. Chinese Premier Wen Jiabao in November 2009 stated a national target to reduce GHG emissions by reducing carbon intensity (emissions per unit of economic output) by 40%-45% by 2020 compared with 2005 levels.[62] The State Council indicated that this carbon-intensity target will be made a "binding goal" in China's 12th Five-Year Plan, from 2011-2015, and long-term national social and economic development plans.

One Chinese researcher has estimated that, if these Chinese economy doubles by 2020, the 40%- 45% target would hold GHG to approximately today's emissions level.[63] The Chinese climate change website suggests that Chinese leaders are "mulling" GHG goals of improvement of carbon intensity of 4%-5% annually over several decades, which could lead to an 85%-90% reduction of carbon intensity by 2050 compared to the 2005 rate.[64] (A percentage improvement expressed as carbon intensity would be easier to achieve than the same percentage target expressed as energy intensity, so this rate of annual improvement would be less than the annual energy intensity improvement target in the current five-year plan.)

2. Principal Policy Instrument(s)

Edicts specify national, provincial, and plant-specific targets or actions. For example, one national goal is to reduce energy consumption per unit of GDP by 20% from 2006-2010. Each province was given a corresponding target in June 2006, and many local governments were assigned energy conservation targets by the National Development and Reform Commission (NDRC) in July 2006. Some of the key instruments the central government is using to meet its targets for 2010 include:

- reducing or eliminating incentives for energy-intensive exports (e.g., export tax rebates);
- implementing a program of "Large Substitute for Small," closing half of small, inefficient electric power plants by 2010, and banning new small plants;
- removing some subsidies from inefficient or polluting plants;
- setting 2010 energy consumption targets within the Top- 1000 Enterprise Program for each large enterprise (in total representing 33% of national energy use in 2004);
- requiring closure of small and inefficient industrial plants, sometimes with compensatory payments;
- setting electricity dispatch rules to favor low-carbon generation, such as feed-in tariffs for renewably produced electricity that can reach 25%-50% higher than coal-based electricity prices;
- providing large subsidies to help finance some large capital investments in efficient or low-emitting technologies;
- allowing energy prices to rise to international price levels in many cases, and imposing (and reportedly beginning to collect) pollution fees;
- setting new vehicle efficiency standards at the Europe-IV level (tighter than U.S.), and making payments to turn in and destroy older, polluting vehicles (like "cash for clunkers");
- raising investments in inter-city and intra-city rail; and
- tightening building efficiency codes by many municipalities, although enforcement may be spotty.

High-level officials have indicated that the 12th Five-Year Plan will embody the -40% to -45% carbon-intensity targets, and that several national laws will be amended in the near-term to achieve GHG reductions. Carbon cap-and-trade "pilot" projects will be initiated in "some designated areas and industries."[65] President Hu has summarized additional targets that likely would help to restrain expected growth of GHG: a target to increase non-fossil fuel share of primary energy consumption to 15% by 2020, and to increase forest coverage by 40 million hectares and forest stock volume by 1.3 billion cubic meters by 2020 from 2005 levels. China also requires strict fuel efficiency standards for vehicles.

Some have argued that China's policies may be undermined by incomplete implementation, due to sometimes vague statement of requirements, lack of enforcement resources, poor data, conflicting priorities at the local level, and other factors. Though some argue that reporting and enforcement of the targets and regulations have been irregular, there are indications that the central government is working to improve such weaknesses, and to impose career penalties on officials who do not meet their targets.[66] The State Council stated in November 2009 that new measures would be developed for auditing, monitoring, and assessing implementation of the GHG plans. Others are cautious about the central government's will and ability to gain full implementation of national policies at the provincial and local levels.

3. Covered Gases and Sectors

Policies are mostly focused on energy reforms not GHG control, though they also reduce CO_2 and methane emissions. Some projects under the Kyoto Protocol's Clean Development Mechanism address many industrial gases (such as hydrofluorocarbons) as well. Sectors addressed include energy, vehicle manufacturing, building, energy-intensive industries, forestry, etc. Agriculture seems engaged only through development of bio-fuels.

4. Allocation of GHG Reductions to Various Sectors

Many sectors are covered through various programs. Targets and actions are set by enterprise, not industry-wide.

5. Any Regulations or Exemptions Specific to Trade-sensitive Sectors

Many Chinese industry-specific policies seem aimed at eliminating the most energy-intensive and inefficient facilities within a sector. Many of China's exporting firms perform close to or at international energy-intensities. In 2007, China removed or reduced export tax rebates for many types of export products, including for energy-intensive, trade-sensitive industries. These adjustments generally have the effect of reducing incentives to export. Examples of additional programs are provided below.

Iron and Steel

The Chinese government has been emphasizing restructuring and improving the overall production efficiency of the iron and steel industry, much of which is likely also to reduce direct and indirect emissions. Closures are mandated in 2006-2010 of 100 million tons of iron production capacity and 55 million tons of steel capacity using inefficient and old technologies.[67] From 2006-2008, 61 million tons of iron and 43 million tons of steel capacity were closed, according to government statistics.[68] Mergers and acquisitions are being encouraged to increase concentration and efficiency in the industry. The adjustment and revitalization plan also envisions shifting the product composition of the sector's production, as well as shifting to integrated capacity.

Aluminum

Chinese requirements for energy savings and emissions reductions in its aluminum industry have been estimated to achieve its target of reducing GHG from the industry by 25% by the end of 2010.[69] The central government mandated closures of inefficient aluminum smelting capacity in 2006-2010. China's Ministry of Finance announced it would levy a 15% export tariff on non-alloy aluminum rods and poles, and eliminate the 5% import duty on electrolytic aluminum and many other energy-intensive commodities, in order to "further restrict exports of high energy-consuming and polluting resources products and encourage imports of raw materials," as well as to suppress China's trade surplus.[70]

The Chinese government has removed preferential electricity rates for metal producers, so manufacturers now pay market prices. The (U.S.-based) Aluminum Association also notes, "Additionally, China has invested in alternative energy systems that will begin paying off in 2009, namely solar and hydroelectric power, which will reduce the cost of energy."[71] This is likely also to reduce associated GHG emissions.

Cement

China set a target to reduce energy intensity in its cement industry by 20% in the 11th Five-Year Plan (2006-2010), using plant closures and installing state-of-the-art technologies. China's cement production is about 50% of the global total. The central government mandated closures of inefficient cement production capacity in 2006-2010, with closures of about 140 million tons of production capacity achieved from 2006-2008.[72] One program is set to "design an economically viable, environmentally friendly alternative fuel and raw materials co-processing program, which will include conducting demonstrations in six Chinese plants, and developing, documenting, and disseminating technical guidelines for co-processing.... [T]ools, training materials, and results from the project will be disseminated to further enhance the capacity building of the entire Chinese cement industry. An integrated national database on energy efficiency and emissions for Chinese cement industry, using worldwide recognized methodologies and tools, will also be established."[73]

Motor Vehicles

New vehicle efficiency standards have been set at the Europe-IV level (stricter than US standards). National policy and investment promotes rail rather than road transport.

China has enacted its version of the "Cash for Clunkers" program: from Aug 1, 2009, to June 30, 2010, consumers may receive 3,000-6,000 Yuan (US$440-875)[74] per vehicle to replace "yellow tag" passenger cars, vans, and trucks that exceed emission standards, or are 8-12 years old. Previous changes in vehicle taxes, with higher rates for large cars and lower rates for small ones, resulted in increased small car sales in 2008.

The total trade-in subsidy, mainly targeting light commercial vehicles, is likely to cost the government around 5 billion Yuan.

INDIA[75]

1. Overall GHG Emission Target, if any, and Timing

The Minister of State for Environment and Forests, Jairam Ramesh, announced on December 3, 2009, that India will reduce its GHG emissions intensity (emissions per unit of GDP) by 20%- 25% by 2020, compared to the 2005 level.[76] (He reportedly said that India's carbon intensity decreased by 17.6% from 1990 to 2002.) Ramesh also committed to India's Parliament that India would accept neither legally binding targets nor peaking dates[77] internationally.[78,79] Earlier, the government also pledged that 20% of India's energy would come from renewable resources by 2020, and 15% of India's annual GHG emissions would be taken up by forests by 2030[80] (up from about 11% in 2005[81]). The Indian government has pledged that its emissions per capita would always remain below those of the now-industrialized countries (though expected population increases are substantial).

2. Principal Policy Instrument(s)

Ramesh has indicated that the national Planning Commission has agreed that India's 12th Five Year Plan, from 2012-2017, will include a low-carbon growth strategy. He identified five categories of measures:[82]

- mandatory fuel efficiency standards for all vehicles by December 2011;
- national building code guidance for energy efficiency, to recommend to local governments to make mandatory;
- amendments to laws to reduce energy intensity of industrial activities;
- forest monitoring; and
- use of advanced technologies (super critical, ultra super critical, and coal gasification) for half of all new coal-fired power plants.

In actions to date, India's national government has relied almost exclusively on public information, training of energy auditors, voluntary "declarations" of energy management policies by businesses, and small financial awards as its principal instruments to promote energy efficiency. In concept, Ramesh has said that India might enact a law directing the government to set climate-related, but non-mandatory, targets, with reporting to and review by the Parliament. He has indicated that the new law may be similar to the Fiscal Responsibility and Budget Management law (FRBM), which directs the government to develop targets, and requires reporting to the Parliament, as well as Parliamentary approval. The targets in the FRBM are neither specified nor binding.

Prime Minister Manmohan Singh approved in August 2009 a national energy efficiency plan that would require 714 energy-intensive industrial facilities in nine sectors, accounting for 40% of India's fossil fuel use, to meet energy efficiency targets. The energy efficiency plan is estimated by 2015 to avoid about 5% of India's projected fossil fuel use. The Prime Minister's Office may be contemplating setting up a new National Climate Change Mitigation Authority under the Prime Minister's authority.

Reportedly, the government has initiated greenhouse gas abatement plans in the past several months, including reforestation. An existing voluntary set of efficiency standards is expected to become mandatory by 2010. Stronger standards may be set for energy efficiency for certain appliances and government buildings; an Energy Conservation Building Code (ECBC) for all new government buildings; and monitoring of afforestation. Prime Minister Singh announced in late August the intention of introducing an energy efficiency trading system to reduce India's energy consumption by 5% and its CO_2 emissions by 100 million tons annually from projected levels by 2015 (about 8% of current emissions).[83] Two funds would be created with about $60 million of funding to provide partial loan guarantees and venture capital. Proposed targets may be set by December 2010.

In 2008, the Prime Minister released a National Action Plan on Climate Change, containing eight "national missions": the National Solar Mission; National Mission for Enhanced Energy Efficiency; National Mission on Sustainable Habitat; National Water Mission; National Mission for Sustaining the Himalayan Ecosystem; National Mission for a Green India; National Mission for Sustainable Agriculture; and National Mission on Strategic Knowledge for Climate Change.[84] The most concrete measures aimed at increasing solar energy capacity. In November 2009, the Indian Union Cabinet approved a Jawaharlal Nehru National Solar Mission (NSM) to increase India's solar electric capacity from 5 megawatts (MW) to 20 gigawatts (GW) by 2022 (slipping back two years from the initial target date), at a cost of $19 billion.[85] Some $900 million has been approved for the initial phase, to install 1.1 GW of on-grid and 0.2 GW of off-grid solar capacity by 2012. The NSM will offer financial incentives to investors, including tax breaks, and will boost research. Several existing laws support renewable energy development, according to a report from the Pew Center.

The Electricity Act (2003) encourages the development of renewable energy by mandating that State Electricity Regulatory Commissions (SERCs) allow connectivity and sale of electricity to any interested person and permit off-grid systems for rural areas. The National Tariff Policy (2006) stipulates that SERCs must purchase a minimum percentage of power from renewable sources, with the specific shares to be determined by each SERC individually. The states of Himachal Pradesh and Tamil Nadu have the highest quotas—20% by 2010 and 10% by 2009, respectively. Under the Rural Electrification Policy (2006) electrification of all villages must be completed by 2012.[86]

India established a program to replace 400 million incandescent light bulbs with efficient compact fluorescents by 2012.

A fund supports the regeneration and sustainable management of forests. The initial capitalization of the fund was proposed to be $2.5 billion, with an annual budget of about $1 billion.[87]

Although India has some pollution control standards in place, enforcement of standards has been low.[88] The current government is planning to establish a new National Environmental Authority,[89] apparently to be modeled after the U.S. EPA.

3. Covered Gases and Sectors

Most identified and proposed measures address CO_2. The proposed system of "tradable energy efficiency certificates" would apply to 714 energy intensive facilities in the following sectors: fossil fuel-fired electricity generation; fertilizer production; cement; iron and steel; chlor-alkali production; aluminum; rail transport; and textiles.

4. Allocation of GHG Reductions to Various Sectors

The Bureau of Energy Efficiency would assign energy efficiency improvement targets to the most energy-intensive industrial plants, based on benchmark performance "bands." Facilities in the most efficient "band" would have a less stringent improvement target, while those in less efficient "bands" would be required to make greater improvements. Facilities that perform better than the targets would receive energy savings certificates ("ESCerts") that could be sold to companies for compliance with their targets or, potentially, banked to meet future requirements. Facilities that fail to meet targets could be fined.

5. Any Regulations or Exemptions Specific to Trade-sensitive Sectors

Reportedly, Indian officials have suggested taxing imports based on the per capita carbon emissions of the exporting country.[90] This could have a large impact on the United States, as its per capita emissions are higher than most countries. (Besides foods and fossil fuels, the United States exports to India a wide variety of products, among which the largest in value are: civilian aircraft and parts, steel and other metal products, synthetic fertilizers, chemicals, electronics and industrial equipment, electronics, and gem diamonds.)[91]

Motor Vehicles

In India, high taxes are levied on motor fuels: 52% on gasoline and 32% on diesel in 2007. The Prime Minister's office has directed the Bureau of Energy Efficiency to set fuel efficiency labeling standards for vehicles under the Energy Conservation Act, to become effective by 2011. However, after several years' delay, these standards have not been set. As planned, the standards would require labeling only by 2011, with mandatory performance to be effective later. The Bureau of Energy Efficiency would certify the manufacturers' labels. Reportedly, some representatives of the automobile sector have demanded that the standards be set on the basis of CO_2 emissions and legally be put on India's list of "local pollutants."[92]

JAPAN[93]

1. Overall GHG Emission Target, if any, and Timing

Under the Kyoto Protocol, Japan agreed to reduce its GHG emissions to 6% below 1990 levels in the period 2008-2012. The Japanese Ministry of

Environment has estimated that national GHG emissions were about 1.9% above its 1990 Kyoto Protocol baseline or almost 8% above its obligation, though this comparison does not account for sequestration or international GHG credits. GHG emissions were 1,286 million tons in the Fiscal Year (FY) 2008-2009, about 6.2% below the previous year, due largely to the economic recession.[94]

In mid-2008, then-Prime Minister Fukuda offered to reduce Japan's GHG by 80% from 2008 levels by 2050, and by 8% below 1990 levels by 2020 (without using international credits). Newly elected Prime Minister Yukio Hatoyama pledged Japan to a GHG target of 25% below 1990 levels by 2020, conditional on all major countries' participation in a new international accord. (The outgoing government's proposed target was equivalent to 8% below 1990 levels. In 2008, Japan's GHG emissions were almost 16% above its Kyoto Protocol target.)

Despite the lack of an internationally binding agreement at the Copenhagen climate conference, Environment Minister Sakihito Ozawa and Minister of Economy, Trade, and Industry Masayuki Naoshima reportedly reiterated the Hatoyama administration's pledge of a 25% emission reduction below 1990 levels by 2020 at a press conference on December 22, 2009. At least one minister has noted that this pledge comes "with conditions" not specified.[95]

2. Principal Policy Instrument(s)

The Japanese Government formulated in 2005 the Kyoto Protocol Target Achievement Plan (KPTAP) to promote measures to cope with global warming. The KPTAP lays out estimated emissions and expected reductions by sector, and for several specific programs, in order for Japan to meet its Kyoto Protocol target. The 2008 review and revision of the plan called for further actions to close the gap between expected emissions and the Kyoto target, including more stringent efficiency standards for equipment, vehicles, and small businesses. The government plan concluded that it would be very difficult to constrain emission reductions associated with the residential and commercial sectors, and therefore relied on expanding the Voluntary Action Plans in the business sector to achieve 80% of the envisaged further GHG reductions.[96] (See section on covered gases and sectors, below.)

Since October 2008, Japan has established an integrated domestic GHG emissions market, comprised of four components: (1) Japan's Voluntary

Emission Trading System (J-VETS) capand-trade system, initiated in 2005 for voluntary trading of CO_2 emissions from energy and process emissions covering only industries that do NOT have in place a Voluntary Action Program; (2) an Experimental Japanese Emissions Trading System, with emissions targets based on industry-specific Voluntary Action Programs; (3) Domestic Credit Scheme, to allow GHG reduction credits (i.e., "offsets") from small and medium-sized companies; and (4) Kyoto Credits, available through any of the three Kyoto Protocol emissions trading mechanisms.

The new Hatoyama government has indicated it plans to create a mandatory GHG cap-and-trade system, require "feed-in" tariffs as financial incentives for renewable energy generation, and may consider a carbon tax.[97] The Hatoyama campaign, on the other hand, pledged before the election to eliminate highway tolls and a fuel tax of about 25 yen (US$0.28)[98] per liter on gasoline by April 2010, which could raise vehicle GHG emissions by as much as 20%.[99]

The Law Concerning the Promotion of Measures to Cope with Global Warming[100] enacted in 1998, directed the national government to promote GHG emission reductions and to enhance carbon sinks. It also directed local governments and business to take actions to limit emissions. This basic authority also directs the central government to publish Japan's GHG emissions.

The 5,000 largest businesses in Japan have been required to report their energy production and consumption for more than a decade by the Law Concerning the Rational Use of Energy.[101] Consequently, the foundation for calculating the energy-related CO_2 emissions from each industrial source is established.

The Act on Promotion of Global Warming Countermeasures and Act on Rational Use of Energy establish authorities to promote energy efficiency in "energy-using" equipment, buildings, factories, and machinery. These and related legislation require efficiency labeling, and allow for low-interest financing, industrial improvement bonds, tax exemptions and other financial incentives to promote efficiency. They also require efficiency measures by industrial facilities and for appliances. The Energy Conservation Center of Japan (ECCJ) is a public-private partnership for research and implementation of energy conservation programs (including Japan's Energy Star program, modeled after the US EPA's), accreditation of energy managers, and information.

3. Covered Gases and Sectors

Under Japan's Kyoto Protocol Target Achievement Plan, industry is expected to reduce its GHG emissions to 7% below 1990 levels during the Kyoto first commitment period (2008-20 12). The Keidanren Voluntary Action Plan[102] on the Environment (VAP) covers 35 industries, include energy, mining, construction, and at least some manufacturing sectors (e.g., production of vehicles, electronics, steel, cement, etc.).

4. Allocation of GHG Reductions to Various Sectors

The Keidanren VAPs include a non-binding target of reducing CO_2 emissions in industry and energy-converting sectors "below" their 1990 levels by 2010. In the Keidanren VAPs, different industries' metrics of performance and targets differ. In 2007, about 18 industries tightened their voluntary targets, although some observers have criticized even the more stringent targets as being no more than what was already being accomplished. Others argue that the voluntary targets are costly compared to reductions expected in other countries, such as within the European Union.

5. Any Regulations or Exemptions Specific to Trade-sensitive Sectors (See Figure 1.)

Motor Vehicles
The Japanese government provides tax benefits for "eco-friendly" vehicles and exemptions from taxes for three years for "next-generation" vehicles.[103] Beginning in April 2009, subsidies have been offered to purchasers of eco-friendly vehicles (e.g., for cars: 100,000 yen, or US$1100). These include a "cash-for-clunkers"-type program that offers higher subsidies to owners who scrap vehicles 13 years or older and replace them with eco-friendly vehicles (e.g., for cars: 250,000 yen, or US$2700). The subsidies extend as well to minivans, trucks and buses. One industry official reported that, with the subsidies, "eco-friendly" vehicles accounted for almost half of vehicle sales in Japan.[104]

Japan is reputed to have among the most stringent fuel economy standards for vehicles in the world, at 46.9 miles per gallon by 2015 (see Appendix). These are expected to constrain new passenger vehicle emissions of GHG.

Iron and Steel

To contribute to Japan's Kyoto Protocol obligations, the Iron and Steel Federation set a voluntary target for the sector of reducing CO_2 emissions by 9% from its 1990-1991 (financial year) levels (200.6 million metric tons) during the period 2008-2012. Due largely to the recession, the industry's emissions were 178.2 million tons in 2008-2009, reflecting a 13% reduction in steel output from the previous year. The industry reportedly also has purchased 56 million tons of GHG reduction credits for delivery during that period.[105] The chairman of Japan's Iron and Steel Federation, Shoji Muneoka, has announced an industry reduction of 5 million metric tons of CO_2-equivalent GHG from their forecast level in 2020. The Federation's businessas-usual projection foresees crude steel production to rise from 2008-2020 by 13%, to 119.7 million metric tons.

KOREA[106]

1. Overall GHG Emission Target, if any, and Timing

On November 17, 2009, the South Korean cabinet approved a 4% GHG emission reduction target by 2020 as a basis for its current and future climate change efforts. The goal is measured from a 2005 baseline and is equivalent to a 30% reduction from "business-as-usual." The target is the most ambitious of three options recommended by the country's Presidential Committee on Green Growth, which had urged South Korea to voluntarily participate in climate change efforts under a midterm target of either an 8% increase, no change, or a 4% cut. President Lee Myung-bak said in a statement released by his office that the decision was made "to facilitate the country's paradigm shift to low-carbon green growth." He characterized the policy as a "voluntary, independent, and domestic target for unilateral reduction," driven by "environmental technology and renewable energy development."[107]

Keidanren Voluntary Action Plan on the Environment (Target and Measures of Major Industry Organizations)

Name of Industry Organization	Target	Measures to Attain Goals
The Federation of Electric Power Companies of Japan (FEPC)	In the period from FY2008 to FY2012, aim to reduce CO_2 emissions intensity (Emission per unit of user end electricity) by an average of approx. 20% or to approx. 0.34kg-CO_2/kWh compared to the FY1990 level.	1. Promotion of nuclear power generation based on security and confidence-building 2. Further improvement of efficiency in thermal power generation and discussion on the management and control of thermal power source 3. Utilization of Kyoto mechanism etc. 4. Development and diffusion of renewable energy
The Japan Iron and Steel Federation (JISF)	Reduce energy consumption in the production process in FY2010 by 10% compared to the FY1990 level on the assumption of the crude steel production 100 million-ton level.	1. Recovery of waste energy (enhancement of TRT, newly installing of CDQ, enhancement of gas recovery, recovery of sensible heat of converter gas, regene-burner, etc.) 2. Efficiency improvement of facilities (introduction of high efficiency oxygen, improvement of electric turbine, improvement of sinter, modification of blast furnace, efficiency of motor, streamlining of power generating facility, modification of hot furnace, etc.) 3. Improvement of operation (reduction of ratio of reduced materials, management of steel temperature, utilization of chilled iron source, etc.) 4. Effective utilization of waste plastic (utilization of waste plastic, enhancement of facility for process of waste plastic, etc.) 5. Others (dust recycle, humidity conditioning for coal, preprocessing of ore, etc)
Japan Chemical Industry Association (JCIA)	1. In the period from FY 2008 to FY 2012, aim to reduce energy intensity to an average of 80% of the FY 1990 level. (However, in case aggravating factors for energy intensity become obvious, it could be about 87%.) 2. Establish guidelines for energy conservation activities in commercial sector, such as headquarters building, sales offices and start the activities. 3. Solicit "Energy conservation activities in residential sector promoted by the chemical industry" which encourages public campaign for energy conservation led by the government from all JCIA members and start the activities. 4. Prepare "Technology handbook for energy conservation and environment of Japanese chemical industry" and provide them to people in developing countries etc. who need energy conservation technologies. 5. Develop and disseminate new materials for energy conservation continuously.	1. Recovery of waste energy (recovery of waste heat and cool energy, turning waste fluid/waste oil/waste gas into fuel, heat storage, etc.) 2. Rationalization of process (process rationalization, process conversion, system change, catalyst change, etc.) 3. Efficiency improvement of facilities and equipment (replacement of equipment and materials, improvement of equipment performance, installation of high efficiency facilities, efficiency improvement of lighting/motor, etc.) 4. Improvement of operation methods (condition change of pressure, temperature, flow, reduction of the number of operating unit, advanced control, reuse/recycle, etc.) 5. Others (product modification, etc)

Name of Industry Organization	Target	Measures to Attain Goals
Petroleum Association of Japan (PAJ)	In the period from FY2008 to FY2012, reduce energy intensity in refineries by an average of 13% compared to the FY1990 level.	1. Revision of operation management (Improvement of control technology and optimization technology) 2. Expansion of mutual utilization of waste heat among facilities 3. Additional construction of recovering facilities of waste heat and waste energy 4. Adoption of efficient equipment and catalyst 5. Efficiency improvement by appropriate maintenance of facilities 6. Participation in "Industrial Complex Renaissance"
Japan Paper Association	In the period from FY2008 to FY2012, aim to reduce fossil energy intensity per product by an average of 20% and CO_2 emissions intensity derived from fossil energy by an average of 16% compared to the FY1990 level. Strive to promote forestation in Japan and overseas to expand owned or managed forested areas to 0.7 million ha by FY2012.	1. Introduction of energy conservation equipment (heat recovery equipment, introduction of inverters etc.) 2. Introduction of high efficiency facilities (high-temperature high-pressure recovery boilers, high-efficiency cleaning equipment, high-dehydrated press, etc.) 3. Revision of manufacturing process (shortening and integration of processes) 4. Fuel switch (switch to biomass energy, waste energy) 5. Strengthening management (review of management value, reduction of dispersion)
Japan Cement Association (JCA)	Reduce energy intensity of cement production (Thermal energy for cement production + Thermal energy for private power generation + Purchased electrical energy) in FY2010 by 3.8% compared to the FY1990 level. * Take an average of five years from FY2008 to FY2012 to achieve the above target.	1. Facilities to utilize waste as alternative heat energy source (waste wood, waste plastic, etc.) 2. Efficiency improvement of facilities (fans, coolers, finishing mills, etc.) 3. Installation of energy conservation equipment (high-efficiency clinker coolers, etc.) 4. Replacement of facilities (including repair of facilities)
Japan Automobile Manufacturers Association, Inc. (JAMA) Japan Auto-Body Industries Association inc. (JABIA) (Two associations integrated their effort to promote the reduction of GHG emission from FY 2007.)	In the period from FY2008 to FY2012, reduce the total CO_2 emissions by an average of 22% compared to the FY1990 level.	1. Energy supply side measures (introduction of cogeneration, improvement of efficiency of boilers, introduction of high-efficiency compressors, introduction of energy conservation facilities) 2. Energy demand side measures (energy conservation in coating line, introduction of invertors for fans and pumps, energy conservation in lighting and air-conditioners, energy-saving operation of compressors, reduction of energy-loss during operation, etc.) 3. Upgrading energy supply methods and technologies of operation and management (reduction of energy loss during no operation, reduction of air leak, etc.) 4. Merger, abolition and integration of lines 5. Fuel switch

Source) Prepared from "Results of the FY2008 Follow-up to the Keidanren Voluntary Action Plan on the Environment (Section on Global Warming Countermeasures, Version Itemized per Business Category), in March 2009" (Website by Keidanren (Japan Federation of Economic Organization))

Note: Table copied from The Energy Conservation Center, Asia Energy Efficiency and Conservation Collaboration Center, 2008. Available at http://www.asiaeec-col.eccj.or.jp/eng/e3104keidanren_plan.pdf.

Figure 1. Japanese Regulations or Exemptions Specific to Trade-Sensitive Sectors

2. Principal Policy Instrument(s)

The November recommendation will empower a governmental committee to prepare industry- specific quotas and implement support measures. Near-term reductions will focus on buildings and transportation to give other industry sectors more time to adjust.

In addition to these recent measures, Korea's policies have involved dialogue with industrial organizations, voluntary plans by participating facilities to save energy and reduce CO_2 emissions, and some non-regulatory emissions trading. The government has provided financial incentives and technological assistance. Voluntary agreements cover plants that consume more than 2,000 tons of oil equivalent annually.[108] This process has resulted in some performance benchmarking for industries, collaborative research, and participation in the Kyoto Protocol's Clean Development Mechanism.

South Korea recently said it plans to invest about 2% of its GDP annually in environment-related and renewable energy industries over the next five years, for a total of US$84.5 billion. The government said it would try to boost South Korea's international market share of "green technology" products to 8% by expanding research and development spending and strengthening industries such as those that produce light-emitting diodes, solar batteries and hybrid cars.[109] To meet its pledge of a new, quantitative target, the government has indicated it may use GHGtrading and tax incentives. It has also indicated that financial incentives would increase use of hybrid cars, renewable and nuclear energy, light-emitting diode lighting, and smart grids.[110]

3. Covered Gases and Sectors

Sectors included in Korea's "Industrial Organization for UNFCCC Task Force Team" are steel, cement, electricity generation, paper, semi-conductor manufacturing, petrochemicals, oil refining, and automobile manufacturing.

4. Allocation of GHG Reductions to Various Sectors

Not yet determined.

5. Any Regulations or Exemptions Specific to Trade-sensitive Sectors

Motor Vehicles

The automobile manufacturing association reached voluntary agreement with the EU to meet CO_2 emission standards of 140grams/km by 2008.[111]

MEXICO[112]

1. Overall GHG Emission Target, if any, and Timing

Mexico voluntarily plans to cut national GHG emissions by 50 million tons per year beginning in 2012, constituting approximately 8% of Mexico's net GHG emissions in 2008. The government has established a non-binding goal to reduce GHG by 50% by 2050 (to 340 million tons of CO_2) below 2000 emissions. The pledge is contingent on availability of international technical and financial support and on successful negotiation of an international agreement consistent with stabilizing CO_2-equivalent concentrations at 450 parts per million. Mexico foresees converging by 2050 on global average emissions per capita at or below 2.8 tons of CO_2 annually.

2. Principal Policy Instrument(s)

In 2007, the Government of Mexico set out a Strategy on Climate Change (NSCC) that identified GHG mitigation opportunities, and vulnerability and adaptation policies. The ensuing Mexico Climate Change Program (MCCP) sets 85 specific goals for mitigating GHG in four emission categories and 12 subcategories. In December 2008, Mexican President Felipe Calderon announced his intention to cap Mexican greenhouse gas emissions and allow GHG trading, beginning with state-owned energy producers. Mexico envisions eventually being part of a domestically regulated but internationally integrated North American GHG trading system.[113]

Mexico mainly promotes energy efficiency (including greater co-generation of heat and power by industrial sources) and renewable energy production, along with prevention of further deforestation, as its mitigation priorities. Principal instruments include Law for the Better Use of Renewable

Energy and the Financing of Energy Transition (2007 or 2008) provide a number of legal energy reforms, including provisions that lay the groundwork for private investment in renewable electricity generation. The Law for the Sustainable Use of Energy created a three-stage program to 2050. It, *inter alia*, promotes renewable energy and energy efficiency. It also requires energy efficiency in all federal, state and local governments.

3. Covered Gases and Sectors

Six Kyoto Protocol gases. The cap-and-trade system under development is likely to cover energy production (oil and gas, refining, electricity), metals, chemicals, textiles, and cement. Analysis is underway to include a cap-and-trade program for vehicle fuel efficiencies as well.

4. Allocation of GHG reductions to various sectors

Not yet determined.

5. Any regulations or Exemptions Specific to Trade-sensitive Sectors

Motor Vehicles
The stringency of Mexico's vehicle efficiency standards was increased in 2004 to a mix of U.S. and European standards for different classes of vehicles.

Oil and Gas Production, Refining and Distribution
PEMEX, Mexico's state-owned petroleum company, has operated an internal carbon cap-and-trade system since 1998.

Russian Federation[114]

1. Overall GHG Emission Target, if any, and Timing

The Russian Federation (hereafter "Russia") projects that its greenhouse gas (GHG) emissions in the year 2010 will be 28% below the 1990 level, which is Russia's GHG emissions cap (its "Assigned Amount") under the Kyoto Protocol.[115] Though GDP in 2006 was 3% below the 1990 level, Russia's GHG emissions were 34% below the 1990 level (inclusive of carbon uptake by forests and other vegetation, net GHG emissions were 74% below the 1990 level). Some four-fifths of the GHG reductions came from the energy sector. Russia's GHG emissions are thus below its Kyoto Protocol obligation, creating a large surplus of emission allowances (Assigned Amount Units, or AAUs, in the terminology of the Protocol). Under the rules of the Kyoto Protocol, Russia may sell its surplus AAUs to other Parties with GHG obligations.

A Presidential Decree[116] on measures for increasing the energy and environmental efficiency of the Russian economy was issued in 2008, setting a target to decrease the energy intensity of the economy by at least 40% by 2020, compared to the 2007 level. The government has also set a target to increase the share of renewable energy (excluding large hydroelectric production) in electricity generation to 4.5% by 2020, and to use 95% of associated natural gas (produced with oil) by 2014-2016.

In the Copenhagen negotiations, President Dmitry Medvedev has offered a GHG target for Russia's emissions of 22%-25% below 1990 levels by 2020.[117] With policies and measures in place, the Russian government has projected that its GHG emissions in 2010, 2015, and 2020 will be reductions of 28%, 21%, and 13%, respectively, of its 1990 emissions level. Other experts project them to be 10%-37% below 1990 levels in 2020 with current policies and economic outlooks.[118]

Although Russian leaders agreed in the G8 summit meeting of July 2008 to consider an 80% reduction from 1990 levels of GHG emissions from developed countries by 2050, Russia leaders agreed only to a 50% reduction target for Russia.

2. Principal Policy Instrument(s)

Many observers contend that climate change has not attracted the interest of high level leaders in Russia and that, consequently, "[t]he government hardly has any official climate strategy, and little progress is occurring."[119] These claims persist in spite of apparent changes in the Russian leadership's diplomatic approach to the issue (e.g., an announcement of a climate "doctrine" accepting that GHG emissions would pose risks and would require actions to reduce emissions).[120] Many suspect that Russia's support for climate change actions is associated with expanding its export market for natural gas in Europe and, to a much smaller degree, the value of potentially selling its surplus AAUs to EU and other countries with GHG reduction obligations.

As noted above, Russia's reduced GHG emissions is due primarily to economic collapse, leading to steep drops in energy demand and production, as well as other activities (e.g., agriculture, waste) that lead to GHG emissions. Replacing old, inefficient manufacturing and other infrastructure has led to relatively slower increases in GHG emissions than in economic activity.

The government's strategy for economic and social development has relied on reform and expansion of the energy sector, in part because 50% of the central government's revenue comes from the oil and natural gas sector.[121] The export value of oil and natural gas has driven a policy emphasizing extraction of these resources for trade. However, many observers have noted a concomitant, low level of investment in new capacity. The 2006 Russian Energy Strategy to 2020 sought to increase reliance on nuclear and coal-fired electricity for domestic use in order to increase oil and natural gas available for export.[122] Investments are being made to back out natural gas use, for example, by investing in efficient, combined cycle gas turbine technologies. These energy initiatives have mixed effects on GHG trajectories.

In 2005, the government adopted the Complex Action Plan for Implementation of the Kyoto Protocol in the Russian Federation for 2004-2008. It gave coordinating authority to the Interdepartmental Commission on Implementation of the Kyoto Protocol in the Russia Federation. It established some sectoral targets for improving energy efficiency, although some commentators allege that no actions would be needed to achieve them.[123] The UNFCCC in-depth review concluded that these targets had been only partially met.

The Mid-term Social-economic Development Programme of the Russian Federation for 2003–2005 provided for economic incentives to modernize equipment and technologies, improving energy efficiency and thereby reducing GHG emissions. To supplement these initiatives, a Presidential Decree was issued in 2008 on measures for increasing the energy and environmental efficiency of the economy of Russia. Other reported actions include:

- Gazprom, Russia's state-owned natural gas enterprise, established an energy conservation program for 2001–2010.
- Gazprom is implementing measures to reduce CH_4 and CO_2 emissions through 2012 (the annual reductions expected are a 10% reduction in CH_4 emissions and a 2.5% reduction in CO_2 emissions); other measures to increase the efficiency of gas transport and decrease losses by Gazprom (emission reductions of 3 Mt CO_2 in the period 2001–2004 through reconstruction of pump stations).
- A federal program for housing for 2002–2010 targets housing retrofit and modernization and includes energy efficiency measures and introduction of small-scale renewable energy generation in the residential and services sectors.

On November 12, 2009, President Medvedev addressed the Federal Assembly and outlined his proposal for Russia to "undergo comprehensive modernization." In this speech Medvedev announced that "increasing energy efficiency and making the transition to a rational resource consumption model is another of our economy's [five] modernization priorities."[124] To this end, he highlighted a number of new program proposals to:

- produce and install individual energy meters for households;
- transition to energy-saving light bulbs from 2011 to 2014;
- introduce energy service contracts and introduce payment for consumption of services (and considering family incomes);
- increase efficiency in the public sector; and
- capture and sell natural gas co-produced with oil, instead of flaring gas.

President Medvedev also promoted developing waste-to-energy systems; super-conductors for electricity production, transmission, and use; and nuclear generation, including nuclear fusion. Some of these proposals were enacted into law in November 2009. The Russian government plans to provide 1.8 trillion rubles ($62.5 billion) for energy-saving projects by 2020.[125] According to the Kremlin website,[126]

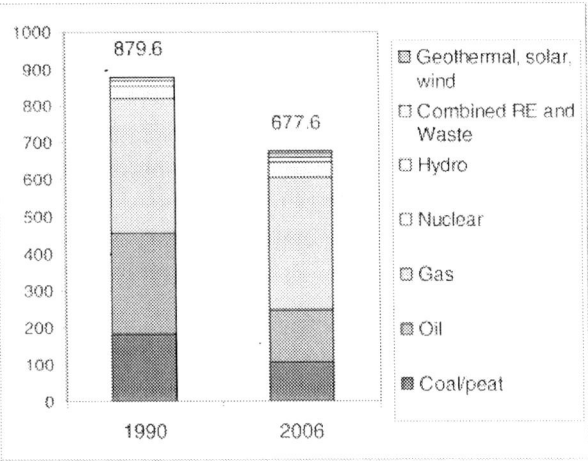

Source: Novikova, 2009, op. cit.

Figure 2. Russian Total Primary Energy Supply, 1990 and 2006 in million tons of oil equivalent (mtoe)

... the new federal law introduces restrictions on the sale of incandescent light bulbs, sets requirements for providing energy efficiency information on goods' labeling, and also brings in provisions on mandatory commercial inventories of energy resources, new buildings' energy efficiency, and reductions in budget spending on purchasing energy resources. The new law also introduces energy evaluations for the most energy-intensive organisations and sets out provisions for transition to long-term tariff regulation and the establishment of a common inter-ministerial energy efficiency information and analysis system.

Some observers have expressed reservations about Russia's implementation of these policies, based on past performance.[127]

The in-depth review of Russia's Fourth National Communication under the United Nations Framework Convention on Climate Change (UNFCCC) found that Russia did not report on its specific domestic measures to abate GHG emissions or detail on how they would contribute to meeting Russia's GHG commitments.[128] The review recommended that the government provide greater transparency of how Russia's policies and measures may be modifying long-term trends in anthropogenic GHG emissions and removals. According to the UNFCCC in-depth review,

In the period 1990–1998, GHG emissions decreased almost in parallel with the economic decline. In the period 1998–2006, GDP growth was accompanied by a relatively slower increase in the level of GHG emissions, which was 9.9 per cent higher in 2006 than in 1998. The differences between GDP and the GHG emission trends are mainly driven by: shifts in the structure of the economy (particularly of non-energy intensive industries); shifts in the primary energy supply (the share of oil and coal has decreased and the share of natural gas and nuclear energy has increased); a decline in activities in the agriculture and transport sectors; the decrease in population (by 3.9 per cent); and the increase in energy efficiency. These trends resulted in a 31.9 per cent decrease in the Party's carbon intensity per GDP unit in 2006 compared with that in 1990.

Russia has not reported estimates of how government funding or financial incentives may influence GHG emissions.

Russia's latest energy strategy, as updated in August 2009, focuses in 2013-2015 on recovery from the current economic crisis. In its second phase, from 2015 to 2022, Russia would emphasize introducing new technologies and more efficiency into its energy sector. An expansion of renewable energy, including large hydroelectric plants, wind, and solar generation, would occur only in the third phase of the new strategy, from 2022 to 2030, along with continued development of hydrocarbon resources.

3. Covered Gases and Sectors

Russia's target under the Kyoto Protocol includes the six Kyoto Protocol gases.

4. Allocation of GHG Reductions to Various Sectors

None specified.

5. Any Regulations or Exemptions Specific to Trade-sensitive Sectors

Motor Vehicles

In 2005, limits on motor vehicle pollutant emissions were introduced, including indicators of GHG emissions. These standards were comparable to the EURO 2–EURO 5 emission standards. (See **Figure A-2** in the **Appendix.**)

UNITED STATES

1. Overall GHG Emission Target, if any, and Timing

The United States has not set legally binding targets to reduce its greenhouse gas (GHG) emissions, neither under domestic law nor international treaty. The House of Representatives passed a bill in June 2009 (H.R. 2454, the American Clean Energy and Security act of 2009) that would cap GHG emissions at about 17% below 1990 emissions by 2020 and 83% below by 2050. The Senate has been working on similar legislation, including S. 1733, the Clean Energy Jobs and American Power Act, which contained a cap of 20% below 1990 levels by 2005 and 83% below by 2050 when it was passed by the Committee on Environment and Public Works in November 2009.

On November 25, 2009, the White House announced that President Obama would attend the December 2009 international negotiations in Copenhagen on an agreement to address climate change beyond the year 2012. The White House stated that he is prepared to offer a "provisional" emissions reduction target of 17% below 2005 levels by 2020, and "ultimately in line with final U.S. energy and climate legislation."[129] On a path consistent with "pending legislation" for a long-term policy objective of 83% below 2005 levels by 2050, U.S. GHG emissions would be 30% and 42% below 2005 levels in 2025 and 2030, respectively, according to the White House.

Had the United States become a Party to the Kyoto Protocol, it would have had an obligation to reduce GHG emissions by 7% below 1990 levels during the first commitment period of 2008- 2012. In 2007, U.S. GHG emissions were about 16% above 1990 levels.[130]

Of the 50 States, 23 have set state-wide GHG mitigation targets, of which six are caps (maxima). While some are enforceable, others are not.

2. Principal Policy Instrument(s)

Current federal climate change policies provide incentives, but few requirements, explicitly to reduce GHG emissions; many programs exist, however, that contribute to limiting GHG emissions through energy efficiency standards, and technical assistance and financial incentives for renewable energy or other low-emitting technologies. For example, a number of tax incentives are in place to encourage investment in renewable energy, more efficient vehicles, and efficiency improvements to buildings. The White House identifies more than $80 billion of funding for clean energy provided under the American Recovery and Reinvestment Act of 2009 (P.L. 111-5), including the "largest-ever investment in renewable energy."[131] Other incentives induce agricultural producers to enhance soil carbon. While temporary financial incentives have been associated with greater investments, some stakeholders have indicated that longer duration of the incentives and combining with other market correction measures are important to effectiveness.

A suite of federal[132] programs, including the Energy Star, Climate Leaders, and Climate Challenge branded initiatives, provides information, technical assistance, and nominal awards to businesses, universities, and other consumers to quantify and reduce their GHG emissions; such programs generally are intended to encourage emission reductions that are already economical but do not occur because of market inefficiencies.

Some GHG reductions are achieved by existing or contemplated regulations. A major regulatory effort governs the energy efficiency of vehicles. For example, Corporate Average Fuel Economy (CAFE) standards will tighten for Model Year 2011 cars and trucks to approximately 27.3 miles per gallon (mpg). Again, these regulations have been put in place for reasons other than abating climate change. However, the Department of Transportation and the Environmental Protection Agency (EPA) are coordinating to propose new, joint CAFE and GHG emission standards for Model Years 2012-2016. The proposal would reach an estimated combined average of 34.1 mpg by 2016 (Table 1); combined with EPA's compliance credits for improving air conditioners of vehicles, the improvement could reach the GHG equivalent of 35.5 mpg. The proposed rules contain flexibilities for manufacturers to comply with the new standards by earning credits by over-complying, or by producing alternative or dual-fueled vehicles. Holders of credits may use them for compliance of other model years or classes, or trade them to another manufacturer. The agencies project that the new standards would reduce GHG emissions by about 900 million metric tons,[133] and reap net cost savings over the lifetimes of vehicles.

The United States has set minimum standards of energy efficiency for a wide variety of residential and commercial equipment since the 1970s, with updates by several more recent laws.[134] Efforts are currently underway to address a backlog of regulations, such as for residential water heaters, dishwashers, clothes dryers, and for commercial motors and lamps, and a number of new, more stringent standards were issues in 2009. About two dozen additional standards are planned over the next few years. In some instances, states may have set appliance efficiency standards more stringent than federal standards (e.g., television standards in California).

Methane emissions from landfills are controlled along with other air pollutants under the Clean Air Act. According to EPA, the regulation requires installation of gas collection and control systems for new and existing landfills and, generally, routing the gas to an energy recovery system. The gas control system must reduce collected landfill gas (LFG) emissions by 98%.[135]

Large programs are devoted to developing new technologies that would be necessary to reduce GHG emissions below current levels. Many experts contend that voluntary efforts (such as the U.S. Climate Leaders Program), research on technologies, and existing regulatory and tax incentives cannot achieve the GHG reductions necessary to avoid "dangerous" climate change.

Of the $6.4 billion in U.S. federal funding in FY2008 for climate change activities, almost all was for scientific and technological research and development. In addition, tax incentives that could help to reduce GHG emissions were equivalent to about $1.5 billion in FY2008. As mentioned above, more than $80 billion in funding was available in FY2009. Funding for regulatory, voluntary, and public education programs was a few percent of the total. President Obama has also pledged, along with leaders of more than 20 other countries, to seek to phase out subsidies for fossil fuels, reducing associated GHG emission by an estimated 10% or more by 2050.[136]

The 110th Congress enacted two broad pieces of legislation—an omnibus energy bill (P.L. 110-140) and a comprehensive appropriations act (P.L. 110-161)—that include climate change provisions. Both statutes increase climate change research efforts, and the energy act requires improvement in vehicle fuel economies, as well as other provisions that would reduce (or sometimes increase) GHG emissions. P.L. 110-161 directs the EPA to develop regulations that establish a mandatory GHG reporting program that applies "above appropriate thresholds in all sectors of the economy."

Table 1. Average Required Fuel Economies under Proposed Standards (in miles per gallon for model year vehicles)

	2012	2013	2014	2015	2016
Passenger Cars	33.6	34.4	35.2	36.4	38.0
Light Trucks	25.0	25.6	26.2	27.1	28.3
Combined	29.8	30.6	31.4	32.6	34.1

Source: National Highway Traffic Safety Administration, "NHTSA and EPA Propose New national Program to Improve Fuel Economy and Reduce Greenhouse Gas Emissions for Passenger Cars and Light Trucks" fact sheet available at http://www.nhtsa.dot.gov/portal/site/nhtsa/menuitem. d0b5a45 b55 bf be 582f57529cdba046a0/.

In the absence of a federal regulatory framework to address U.S. GHG emission reductions, a majority of states have established formal GHG mitigation policies, including targets for future reductions. Sixteen states[137] are regulating CO_2 emissions from electric utilities: 11 using a sectoral cap-and-trade approach, and five using emission performance standards. In several regions, including the Northeast, the Midwest and the West, states are working together to create regional schemes to cap GHG emissions and allow trading of emissions permits across borders. All states but four now support "net-metering" to allow producers of renewably generated electricity to sell what they don't use into the electric grid. Twenty-six states have set renewable portfolio standards and another four have set alternative energy portfolio standards; these standards require that a specified share of the state's electricity must be generated by renewable or alternative energy sources by a given date. An additional five states encourage renewable or alternative energy sources with non-binding goals.

In the transportation sector, 15 states, led by California, are adopting GHG emission standards for motor vehicles, and three additional states are poised to follow. Thirty-eight states offer tax exemptions, credits, and/or grants to promote biofuels, of which 13 have set regulations requiring a specified share of motor fuels to come from biomass. To address growth of traffic, 18 states have set "smart growth" policies. Arizona, for example, has enacted laws and required improved coordination of state agency spending to help communities address a variety of growth pressures. Three of these states have also set targets to reduce vehicle miles traveled in the state. For example, the State of Washington set a goal in 2008 to reduce annual per capita vehicle miles traveled by 18% by 2020, 30% by 2035, and 50% by 2050, compared to 1990 levels.

Building codes typically fall under local authorities, although a growing number of states have set performance standards that help to limit GHG emissions. Most states have set efficiency standards for state, commercial, and residential buildings. Twelve have set appliance efficiency standards as well.

Over the past five years, a proliferation of litigation relating to climate change also presses the federal government toward actions to reduce GHG emissions. For example, the Supreme Court ruled in 2007 that the EPA must consider regulating CO_2 and other GHG emitted from motor vehicles as pollutants under the Clean Air Act.[138] The Obama Administration has made clear that it would prefer Congress to enact GHG-specific legislation but that it will move to regulate in the absence of such new law. Further litigation has been pursued, challenging the Executive Branch to action, using the Endangered Species Act, the Energy Policy and Conservation Act and the Outer Continental Shelf Lands Act. A few international-law claims have been filed against the United States as well.[139]

3. Covered Gases and Sectors

Only methane emissions currently are regulated directly, although CO_2 has been proposed to be regulated from motor vehicles (in a joint rule with fuel economy standards) and is reduced through other regulatory measures.

4. Allocation of GHG Reductions to Various Sectors

Because no economy-wide reduction strategy is in place, there is no allocation among sectors.

5. Any Regulations or Exemptions Specific to Trade-Sensitive Sectors

Because no economy-wide reduction strategy is in place, there are no regulations or exemptions in place specific to trade-sensitive sectors. H.R. 2454, which passed the House on June 26, 2009, includes two strategies to address possible shifts of GHG emissions from the United States to less regulated companies in other countries: (1) free allocation of allowances (similar to that of the EU), and (2) an international reserve allowance (IRA) scheme. The scheme would require importers of energy-intensive products from countries with insufficient carbon policies to submit a prescribed amount of "international reserve allowances," or IRAs, for their products to gain entry into the United States. Based on the GHG emissions generated in the production process, IRAs would be submitted on a per-unit basis for each category of covered goods from a covered country. Specifically, H.R. 2454 Section 768 requires EPA to promulgate rules establishing an international reserve allowance system for covered goods from the eligible industrial sector, including allowance trading, banking, pricing, and submission requirements. (See also the Appendix, comparing U.S. efficiency standards for motor vehicles with those of other countries.)

APPENDIX. COMPARISON OF VEHICLE EFFICIENCY STANDARDS INTERNATIONALLY (AS OF MID-2009)

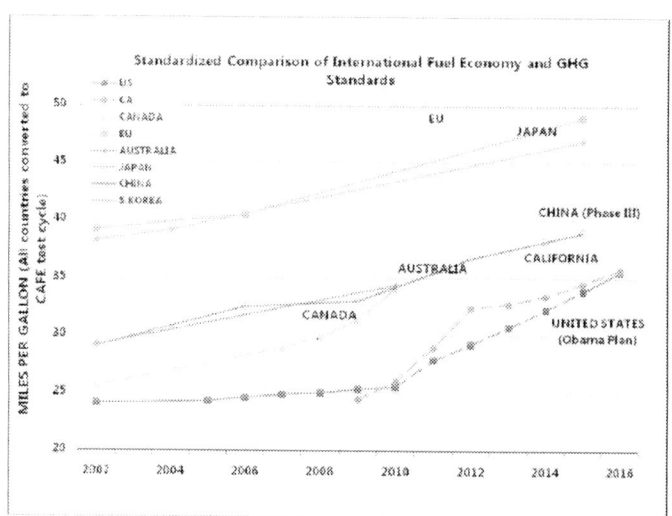

Source: Feng An, "Revised Chart for World Standards," Innovation Center for Energy and Transportation (iCET) (2009). Available at http://www.icet.org.cn.

Figure A-1. Comparison of International Fuel Economy and GHG Standards

CAFE mpg	UNITED STATES	CALIFORNIA	CANADA	EUROPE	AUSTRALIA	JAPAN	CHINA
2002	24.1		25.5	39.2	29.1	38.2	29.0
2003							
2004						39.1	
2005	24.3						
2006	24.6			40.5			32.4
2007	24.9		28.8				
2008	25.0		29.8				
2009	25.3	24.4	31.3				33.0
2010	25.5	26.0	34.1		34.4		
2011	27.8	28.9					
2012	29.2	32.4		48.9			36.7
2013	30.7	32.7					37.4
2014	32.2	33.4					38.1
2015	33.8	34.5				46.9	38.9
2016	35.5	35.7					

Note: all countries/regions normalized to US CAFE driving test cycle and converted to miles per gallon

Source: Feng An, "Revised Chart for World Standards," Innovation Center for Energy and Transportation (iCET) (2009). Available at http://www.icet.org.cn.

Figure A-2. Standardized Comparison of Select Vehicle Efficiency Standards Internationally (standards as of mid-2009)

End Notes

[1] For a further discussion on trade-sensitivity issues, see CRS Report R40100, *"Carbon Leakage" and Trade: Issues and Approaches*, by Larry Parker and John Blodgett; and CRS Report R40914, *Climate Change: EU and Proposed U.S. Approaches to Carbon Leakage and WTO Implications*, by Larry Parker and Jeanne J. Grimmett.

[2] This section was prepared by Richard K. Lattanzio, Analyst in Environmental Policy (7-1754), with input from Larry Parker, Specialist in Energy and Environmental Policy (7-7238) and Jane A. Leggett, Specialist in Environmental and Energy Policy (7-9525).

[3] http://europa.eu/rapid/pressReleasesAction.do? reference=IP/09/1703&format = HTML &aged= 0&language=EN& guiLanguage=en.

[4] See {COM(2008) 13 final}; {COM(2008) 16 final}; {COM(2008) 17 final}; {COM(2008) 18 final}; {COM(2008) 19 final} at http://ec.europa.eu/environment/climat/emission pdf/com_2008_16_en.pdf.

[5] (2003/87/EC); see http://europa.eu/rapid/pressReleases Action.do?reference=IP/09/628& format = HTML&aged=0& language=EN&guiLanguage=en. Also see CRS Report RL34150, *Climate Change and the EU Emissions Trading Scheme (ETS): Kyoto and Beyond*, by Larry Parker.

[6] Each Member State is responsible for the implementation of Community law (adoption of implementing measures before a specified deadline, conformity and correct application) within its own legal system. Under the Treaties (Article 226 of the EC Treaty; Article 141 of the Euratom Treaty), the Commission of the European Communities is responsible for ensuring that Community law is correctly applied. Consequently, where a Member State fails to comply with Community law, the Commission has powers of its own (action for

non-compliance) to try to bring the infringement to an end and, where necessary, may refer the case to the European Court of Justice. For additional information, see http://ec.europa.eu/community_law/infringements/infringements_en.htm

[7] Directive 2009/28/EC of the European Parliament and of the Council of 23 April 2009.

[8] See, for example, Andres Cala, Europe Warming to Carbon Tax, Energy Tribune. "Spain and Ireland, which until recently were considered unlikely candidates to follow suit because of their high unemployment rates, are also weighing adding similar levies next year. Ireland's Finance Minister, Brian Lenihan, said recently that the government would not raise taxes to finance next year's budget, with the single exception of a carbon tax.... Spain's Prime Minister Jose Luis Rodriguez Zapatero, which has announced a fiscal reform to raise more money to control a rampant deficit, called the carbon tax an 'interesting' proposal and added carbon taxes will inevitably be applied by most countries." 23 Sept. 2009. http://www.energytribune.com/articles.cfm?aid=2354

[9] http://news.bna.com/deln/DELNWB/split_ display.adp?fedfid=15354499& vname= dennot allissues &fn=15354499& jd=a0c0y8h5r1&split=0; http://www.reuters.com/article/GCA-GreenBusiness/idUSTRE59544A20091006.

[10] See CRS Report R40090, *Aviation and Climate Change*, by James E. McCarthy.

[11] http://www.consilium.europa.eu/uedocs/cms_data/docs/pressdata/en/misc/107136.pdf.

[12] Directive 2009/33/EC of the European Parliament and of the Council of 23 April 2009.

[13] http://www.consilium.europa.eu/uedocs/cms_data/docs/pressdata/en/misc/107136.pdf.

[14] If one or more countries requires carbon controls that add to production costs in businesses that compete internationally, it is possible for "carbon leakage" to occur if production in the controlled countries declines because purchasers instead buy increased supply from uncontrolled producers in other countries. Though emissions may decline from the controlled facilities, they may increase at uncontrolled facilities, thereby leading to "carbon leakage." This would offset the benefits of the emission controls.

[15] Of the 146 sectors, 117 have trade intensity > 30%; 27 have both estimated CO_2 costs >5% and trade intensity > 10%; and two sectors have CO_2 cost above 30% and trade intensity < 10%. Hans Bergman, "Sectors Deemed to be Exposed to a Significant Risk of Carbon Leakage—Outcome of the Assessment" presentation to Working Group 3 Meeting, 18 September 2009.

[16] http://ec.europa.eu/environment/climat/emission/carbon_en.htm.

[17] This section was prepared by Richard K. Lattanzio, Analyst in Environmental Policy (7-1754).

[18] Live market currency exchange rate for November 19, 2009, is listed as 1 Euro equivalent to 1.49 US$ (http://www.xe.com/). Currency rates are subject to fluctuation.

[19] Economist, Sept. 17 2009. http://www.economist.com/ world/europe/ displaystory. cfm? story_id=14460346.

[20] In the international negotiations held in Copenhagen in December 2010, France (nor the EU) agreed with the United States, Australia, and China to reaffirm a principle not to hide trade protectionism behind climate change policy measures, according to a *New York Times* article. http://www.nytimes.com/2009/12/16/business/global/16trade.html? fta=y.

[21] http://news.yahoo.com/s/afp/20090918/sc_afp/francegermanyclimateenvironmentuneu.

[22] This section was prepared by Richard K. Lattanzio, Analyst in Environmental Policy (7-1754).

[23] http://www.bmu.de/english/climate/doc/39945.php.

[24] See article at http://afp.google.com/article/ALeqM5jRYO-p98IjJ1mzuQxZoS4LODTsMg.

[25] See article at http://www.eubusiness.com/news-eu/1200576720.98.

[26] This section was prepared by Richard K. Lattanzio, Analyst in Environmental Policy (7-1754).

[27] http://www.opsi.gov.uk/acts/acts2008/ukpga_20080027_en_2#pt1-pb2-11g4

[28] For this and other policy descriptions, see the Department of Energy and Climate Change website: http://www.decc.gov.uk/en/content/cms/ publications/lc_trans_ plan/lc_ trans_ plan.aspx

[29] http://www.carbonreductioncommitment.info/carbon-reduction-commitment

[30] Live market currency exchange rate for November 19, 2009, is listed as 1UK£ equivalent to 1.67 US$ (http://www.xe.com/). Currency rates are subject to fluctuation.
[31] http://www.energysavingtrust.org.uk/Global-Data/Funding-Information/Carbon-Emissions-Reduction - Target-CERT.
[32] http://www.publications.parliament.uk/pa/cm200708/cmselect/cmenvaud/590/59003.htm.
[33] The Climate Change Levy revised and replaced a fossil fuel levy.
[34] Maria Pender, "UK Climate Change Programme: Business and Public Sector Economic Agreements."
[35] http://www.berr.gov.uk/energy/environment/euets/index.html.
[36] DECC, *Investing in a Low Carbon Britain*, available at http://interactive.bis.gov.uk/lowcarbon/vision/.
[37] This section was prepared by Jane A. Leggett, Specialist in Environmental and Energy Policy (7-9525).
[38] http://www.environment.gov.au/minister/wong/2009/mr20090504.html.
[39] Live market currency exchange rate for November 19, 2009, is listed as 1Aus$ equivalent to 0.92 US$ (http://www.xe.com/). Currency rates are subject to fluctuation.
[40] Renewable Energy (Electricity) Amendment Act 2009, No. 78, 2009, C2009A00078; and Renewable Energy (Electricity) (Charge) Amendment Act 2009, No. 79, 2009, C2009A00079. http://www.comlaw.gov.au/comlaw/ Legislation/Act1.nsf/0/94CB90B9EED48B69CA25762D001B6F5F?OpenDocument.
[41] Australian Government, Carbon Pollution Reduction Scheme: Australia's Low Pollution Future: White Paper (December 2008).
[42] http://www.climatechange.gov.au/whitepaper/summary/index.html.
[43] Reuters, "UPDATE 1-Australia Govt Secures Carbon Deal with Opposition," November 24, 2009, http://in.reuters.com/articlePrint?articleId=INSYD51352320091123.
[44] http://www.environment.gov.au/minister/wong/2009/mr20090504a.html.
[45] Renewable Energy (Electricity) Amendment Act 2009, Schedule 2.
[46] This section was prepared by Richard K. Lattanzio, Analyst in Environmental Policy (7-1754).
[47] See http://www.reuters.com/article/marketsNews/idUSN1347815120091113.
[48] According to Brazil's National Institute of Space Research (INPE), Brazil's average rate of deforestation from 1996 to 2005 was 7,542 square miles annually, compared to averages of 6,574 annually from 1988 to 1995, and 4,974 from 2006 to 2008; http://www.mongabay.com/brazil.html. This target does not appear to include forests, including open canopy forests, in other parts of Brazil, which may be cleared for agricultural production. Also, *http://en.cop15.dk/* news/view+news?newsid=2351, http://www.cmcc.it:8008/cmcc/blog-en/brazil-sets-new-deforestation-target.
[49] http://www.grist.org/article/2009-12-30-brazils-lula-signs-law-cutting-co2-emissions.
[50] http://www.mma.gov.br/estruturas/208/_arquivos/national_plan_208.pdf.
[51] Brazil received $100 million of the pledge on March 25, 2009. The remainder is pending. See http://inter.bndes.gov.br/english/news/not036_09.asp.
[52] See http://www.eenews.net/Greenwire/2009/11/13/4.
[53] http://www.redd-monitor.org/2009/01/23/brazils-national-plan-on-climate fund- %E2%80%9C this-plan-does-not-create-any-carbon-credits-or-right-to-emissions%E2%80%9D/.
[54] This section was prepared by Carl Ek, Specialist in International Relations (7-7286).
[55] Canada's New Government Announces Mandatory Industrial Targets to Tackle Climate Change and Reduce Air Pollution. News release. Environment Canada website. April 27, 2007. http://www.ec.gc.ca/default.asp?lang=En&n= 714D9AAE-1&news=4F2292E9-3EFF-48D3-A7E4-CEFA05D70C21.
[56] No Clear Environmental Champion; Canada and the United States Have Shown Varied Levels of Aggressiveness in the Fight to Combat Climate Change. *Globe and Mail*. July 9, 2008. See also: Canada's Greenhouse Emissions Soaring Again: UN Report. *Canwest News Service*. April 21, 2009.

[57] Government Delivers Details of Greenhouse Gas Regulatory Framework. News release. Environment Canada website. March 10, 2008. http://www.ec.gc.ca/default.asp?lang=En&n=714D9AAE-1&news=B2B42466-B768-424C-9A5B-6D59C2AE1C36.

[58] Notes for an address by the Honourable Jim Prentice, P.C., Q.C., M.P.: Minister of the Environment on Canada's climate change plan. Speech. Environment Canada website. June 4, 2009. http://www.ec.gc.ca/default.asp?lang=En& n=6F2DE1CA-1&news=400A4566-DA85-4A0C-B9F4-BABE2DF555C7.

[59] CRS discussion with Canadian government official, September 10, 2009.

[60] Notes For an Address by the Honourable Jim Prentice, P.C., Q.C., M.P. Minister of the Environment on New Regulations To Limit Greenhouse Gas Emissions. Speech. Environment Canada website. April 1, 2009. http://www.ec.gc.ca/default.asp?lang=En&n=6F2DE1CA-1&news=D8C4903B-B406-4B70-8A4A-EDEF99B71D38.

[61] This section was prepared by Jane A. Leggett, Specialist in Environmental and Energy Policy (7-9525).

[62] Xinhuanet, "China Announces Targets on Carbon Dioxide Emission Cuts," November 26, 2009, http://www.ccchina.gov.cn/en/NewsInfo.asp?NewsId=20831.

[63] Xinhuanet, ibid.

[64] http://www.ccchina.gov.cn/en/NewsInfo.asp?NewsId=20325. This article also points to a study indicating that an 83% reduction of carbon intensity by 2050 would cost about 2.3% of GDP, while a 90% reduction of carbon intensity would cost about 7% of GDP. It is unclear whether this is a lost compared to the annual rate of GDP growth, or to cumulative GDP growth in 2050.

[65] Jing Li and Zhe Zhu, "Legislature Takes Urgent Action in Climate Change Fight," *China Daily*, August 28, 2009, http://www.chinadaily.com.cn/china/2009-08/28/content_8626140.htm.

[66] See, for example, http://news.xinhuanet.com/english/2007-06/20/content_6269732.htm; http://www.chinacsr.com/en/ 2009/06/18/5487-china-first-heavy-industries-fined-for-infringement-of-environmental-rules/; http://www.china.org.cn/ environment/2009-09/28/content_18619189.htm; and http://www.china.org.cn/government/news/2008-03/12/content_12338958.htm.

[67] http://www.reportbuyer.com/industry_manufacturing china_ steel_ industry.html.

[68] http://news.xinhuanet.com/english/2009-08/25/content_11942981_1.htm.

[69] Feng Gao et al., "Greenhouse gas emissions and reduction potential of primary aluminum production in China," *Science in China Series E: Technological Sciences* 52, no. 8 (2009): 2161-2166, doi:10.1007/s11431-009-0165-6.

[70] http://experts.e-to-china.com/analysis/general_analysis/Taxation/2009/0728/58804.html.

[71] http://www.aluminum.org/AM/Template.cfm?Section=Home&CONTENTID=27780& TEMPLATE=/CM/ ContentDisplay.cfm.

[72] http://news.xinhuanet.com/english/2009-08/25/content_11942981_1.htm.

[73] http://china.lbl.gov/news/chinese-cement-companies-reduce-their-carbon-footprint.

[74] Live market currency exchange rate for November 19, 2009, is listed as 1 CNY = 0.146 US$ (http://www.xe.com/). Currency rates are subject to fluctuation.

[75] This section was prepared by Jane A. Leggett, Specialist in Environmental and Energy Policy (7-9525).

[76] "India's 2020 Target: Reduce Emission by 20-25%," *The Times of India*, December 3, 2009, online edition, http://timesofindia.indiatimes.com/india/Indias-2020-target-Reduce-emission-by-20-25/articleshow/5297073.cms.

[77] A date by which its national emissions would peak and then begin to decline in absolute terms. Some proposals have advocated peaking dates for developing countries of between 2015 and 2030.

[78] T.K. Arun, "For a Binding Climate Target," *The Economic Times (India)*, December 4, 2009, sec. Op-Ed, http://economictimes.indiatimes.com/opinion/columnists/t-k-arun/For-a-binding-climate 5298331.cms.

[79] In the same speech to Parliament, Ramesh stated that India would not allow international review of GHG reduction actions it takes without international financing, though the government "can consider" international review of actions that are supported by international finance.

[80] http://www.forbes.com/2009/09/23/jairam-ramesh-india-business-energy-climate-change.html.

[81] http://www.thaindian.com/newsportal/india-news/indian-forests-absorb-11-of-annual-greenhouse-gas-emissionsjairam-ramesh_100240011.html.

[82] *Times of India*, December 3, 2009, op. cit.

[83] See, for example, http://in.reuters.com/article/oilRpt/idINDEL15998520090907? Page Number =1& virtualBrandChannel=0.

[84] http://www.indg.in/rural-energy/environment/national-action-plan-on-climate-change.

[85] Ministry of New and Renewable Energy, "Statement of Dr. Farooq Abdullah on Jawaharial Nehru National Solar Mission – 'Solar India'" November 23, 3009.

[86] Pew Center, "Climate Change Mitigation Measures in India," International Brief 2, September 2008.

[87] http://online.wsj.com/article/SB125018657071529801.html.

[88] Among many sources: http://www.business-standard.com/india/news/23-thermal-plants 01/09/69289/on.

[89] http://www.business-standard.com/india/news/govt-to-reduce-water-air-pollution/365976/.

[90] http://www.dw-world.de/dw/article/0,,4707051,00.html.

[91] U.S. Census Bureau, Foreign Trade Statistics, http://www.census.gov/foreign-trade/statistics c5330.html.

[92] http://www.greencarcongress.com/2009/06/india-fe-20090603.html.

[93] This section was prepared by Jane A. Leggett, Specialist in Environmental and Energy Policy (7-9525).

[94] National Institute for Environmental Studies, http://www.nies.go.jp/whatsnew/ 2009/2009 1111/20091111-e.html.

[95] BNA, "Japan Stands by Pledge to Cut Emissions 25 Percent by 2020; Industry Voices Dissent," *Daily Environment Report*, 246 DEN A-3, December 29, 2009.

[96] For a summary of the plan in English, see http://eneken.ieej.or.jp/data/en/data/pdf/443.pdf.

[97] Various press reports, including http://search.japantimes.co.jp/cgi-bin/ed20090925a1.html.

[98] Live market currency exchange rate for November 19, 2009, is listed as 1 JPY = 0.0112 USD. (http://www.xe.com/). Currency rates are subject to fluctuation.

[99] http://www.planetark.com/enviro-news/item/54691.

[100] Law No.117 of 1998.

[101] 22 June 1979, Law No. 49. Revised in 10 December 1983, 31 March 1993, 12 November 1993, 9 April 1997, and 5 June 1998.

[102] Established by Nippon Keidanren, the Japan Business Federation. Negotiated environmental agreements in Japan have been used in lieu of legally binding regulation since the 1990s, and are not comparable to "voluntary programs" in the United States or some other countries. For example, they may require inspections and there are few reported instances of non-compliance with set targets (Imura Hidefuri, "Building a Cooperative Relationship Between Industry and Regulatory Authorities," presented at OECD, Environmental Compliance Assurance: Trends and Good Practices Paris, 17-18 November 2008."

[103] http://www.jama-english.jp/asia/news/2009/vol36/index.html.

[104] Ibid.

[105] http://steelguru.com/news/international_news/MTIxMjIw/Japan_steelmakers_ to_receive_ 56_million_tonnes_of_CO2_offsets.html.

[106] This section was prepared by Jane A. Leggett, Specialist in Environmental and Energy Policy (7-9525).

[107] http://www.korea.net/News/News/newsView.asp? serial_no=20091118002& part=101&SearchDay=&page=1.
[108] http://www.wwf.or.jp/activity/climate/lib/kyotoprotocol/20040928b.pdf.
[109] Mufson, "Asian Nations Could Outpace U.S. in Developing Clean Energy," *The Washington Post*, http://www.washingtonpost.com/wp-dyn/content/article/2009/07/15/AR2009071503731.html.
[110] Various press reports, including http://www.reuters.com/article/environmentNews/idUSTRE57308M20090804.
[111] http://www.wwf.or.jp/activity/climate/lib/kyotoprotocol/20040928b.pdf.
[112] This section was prepared by Jane A. Leggett, Specialist in Environmental and Energy Policy (7-9525).
[113] North American Leaders' Declaration on Climate Change and Clean Energy, August 10, 2009. Available at http://pm.gc.ca/eng/media.asp?category=5&id=2724.
[114] This section was prepared by Jane A. Leggett, Specialist in Environmental and Energy Policy (7-9525).
[115] United Nations Framework Convention on Climate Change, *Report of the Centralized In-Depth Review of the Fourth National Communication of the Russian Federation* (Bonn, August 31, 2009), http://unfccc.int/documentation/documents/advanced_search/items/3594.php?rec=j&priref=600005423.
[116] Decree 889, June 4, 2008.
[117] http://www.reuters.com/article/idUSTRE5AH21E20091118?feedType=RSS&feedName=environmentNews&utm_source=feedburner&utm_medium=feed&utm_campaign=Feed%253A+reuters%252Fenvironment+%2528News+%252F+US+%252F+Environment%2529.
[118] See Table 5 in Aleksandra Novikova, Anna Korppoo, and Maria Sharmina, *Russian Pledge vs. Business-As-Usual: Implementing Energy Efficiency Policies Can Curb Carbon Emissions* (The Finnish Institute of International Affairs, December 4, 2009), http://www.upi-fiia.fi/en/publication/97/.
[119] Anne Karin Saether, "Moscow Environmental Conference Places Climate Demands on Medvedev," Bellona, March 27, 2009, http://www.bellona.org/articles/articles_2009/environmentalists_put_climate_changes_to_medvedev; Simon Shuster, "Russia offers climate goal with no real bite," June 19, 2009, http://www.reuters.com/article/environmentNews/idUSTRE55I3CP20090619; Ulkopoliittinen instituuttin, "Russia's Post-2012 Climate Politics in the Context of Economic Growth," May 11, 2008, http://www.upi-fiia.fi/fi/event/195/; or, Simon Shuster, "Russia Still Dragging Its Feet on Climate Change," Time, October 8, 2009, http://www.time.com/time/specials/packages/article/0,28804,1929071_1929070_1934785,00.html.
[120] Quirin Schiermeier, "Russia makes major shift in climate policy," *Nature -News* (May 26, 2009), http://www.nature.com/news/2009/090526/full/news.2009.506.html; Simon Shuster, "Russia offers climate goal with no real bite," June 19, 2009, http://www.reuters.com/article/environmentNews/idUSTRE55I3CP20090619; or 1. Oleg Shchedrov, "Russia's Medvedev warns of climate catastrophe," November 16, 2009, http://www.reuters.com/article/environmentNews/idUSTRE5AF1SU20091116.
[121] Jean Foglizzo, "Russia's New Energy Strategy Seems a Lot Like its Old One," The New York Times, March 30, 2008, http://www.nytimes.com/2008/03/30/business/worldbusiness/30iht-rnrgruss.1.11526942.html.
[122] Kevin Rosner, "Dirty Hands: Russian Coal, GHG Emissions & European Gas Demand," Journal of Energy Security (August 27, 2009), http://www.ensec.org/index.php?option=com_content&view=article&id=207:dirty-hands-russiacoal-ghg-emissions-aamp-european-gas-demand&catid=98: issuecontent0809& Itemid=349. The author raises, "The significant issue is whether it would be more advantageous, from an environmental-security perspective within the framework of Russia's coal paradigm, that

the majority of new coal capacity is driven by comparatively more regulated OECD countries or whether it will revert back to Russia. Russia's environmental record is not exemplary in this regard."

[123] Ibid.
[124] Dimtry Medvedev, "Presidential Address to the Federal Assembly of the Russian Federation,"http://www.kremlin.ru, November 12, 2009.
[125] Sergei Blagov, "Russia Seeks to Sustain its Energy Security," Eurasia Daily Monitor, December 2, 2009,http://www.cdi.org/Russia/johnson/2009-222-20.cfm.
[126] Kremlin, November 23, 2009, http://eng.kremlin.ru/text/news/2009/11/222959.shtml.
[127] For example, Novikova, 2009, op. cit. and Blagov, 2009, op. cit.
[128] UNFCCC, op. cit., p. 4.
[129] White House, "Combating Climate Change at Home and Around the World," November 25, 2009, http://www.whitehouse.gov/blog/2009/11/25/combating-climate-change-home-and-around-world.
[130] United States Environmental Protection Agency, *The U.S. Inventory of Greenhouse Gas Emissions and Sinks: 1990-2007*, EPA 430-F-06-010 (Washington DC: Office of Atmospheric Programs, 2009).
[131] White House, 2009, op. cit.
[132] See http://www.epa.gov/climatechange/policy/neartermghgreduction.html, http://www.pi.energy.gov/, and http://www.usda.gov/oce/climate_change/index.htm.
[133] White House, 2009, op. cit.
[134] Established by Part B of Title III of the Energy Policy and Conservation Act (EPCA), P.L. 94-163, as amended by the National Energy Conservation Policy Act, P.L. 95-619, by the National Appliance Energy Conservation Act, P.L. 100-12, by the National Appliance Energy Conservation Amendments of 1988, P.L. 100-357, and by the Energy Policy Act of 1992, P.L. 102-486, and by the Energy Policy of 2005, P.L. 109-58.
[135] http://www.epa.gov/reg3artd/airregulations/ap22/landfil2.htm.
[136] White House, 2009, op. cit.
[137] Data on state policies come from the Pew Center on Global Climate Change website, extracted November 20, 2009. http://www.pewclimate.org/states-regions.
[138] *Massachusetts v. EPA*, 127 S. Ct. 1438 (2007).
[139] See CRS Report RL32764, *Climate Change Litigation: A Survey*, by Robert Meltz.

In: Countries and Climate Policies and Paths... ISBN: 978-1-61728-923-1
Editors: Thomas P. Parker © 2011 Nova Science Publishers, Inc.

Chapter 2

CLIMATE CHANGE AND EU EMISSIONS TRADING SCHEME (ETS): LOOKING TO 2020

Larry Parker

SUMMARY

The European Union's (EU) Emissions Trading Scheme (ETS) is a cornerstone of the EU's efforts to meet its obligation under the Kyoto Protocol. It covers more than 10,000 energy intensive facilities across the 27 EU Member countries; covered entities emit about 45% of the EU's carbon dioxide emissions. A "Phase 1" trading period began January 1, 2005. A second, Phase 2, trading period began in 2008, covering the period of the Kyoto Protocol. A Phase 3 will begin in 2013 designed to reduce emissions by 21% from 2005 levels.

Several positive results from the Phase 1 "learning by doing" exercise assisted the ETS in making the Phase 2 process run more smoothly, including: (1) greatly improving emissions data, (2) encouraging development of the Kyoto Protocol's project-based mechanisms—Clean Development Mechanism (CDM) and Joint Implementation (JI), and (3) influencing corporate behavior to begin pricing in the value of allowances in decision-making, particularly in the electric utility sector.

However, several issues that arose during the first phase were not resolved as the ETS moved into Phase 2, including allocation schemes and new entrant reserves, and others. A more comprehensive and coordinated response by the EU has been made for Phase 3 with harmonized and coordinated rules being developed by the European Commission.

The United States is not a party to the Kyoto Protocol. However, five years of carbon emissions trading has given the EU valuable experience in designing and operating a greenhouse gas trading system. This experience may provide some insight into cap-and-trade design issues currently being debated in the United States.

- The U.S. requires only electric utilities to monitor CO_2. The EU-ETS experience suggests that expanding similar requirements to all facilities covered under a cap-and-trade scheme would be pivotal for developing allocation systems, reduction targets, and enforcement provisions.
- In the U.S. debate continues on comprehensive versus sector-specific reduction programs; the EU-ETS experience suggests that adding sectors to a trading scheme once established may be a slow, contentious process.
- As with most EU industries, most U.S. industry groups either oppose auctions outright or want them to be supplemental to a base free allocation. The EU-ETS experience suggests Congress may want to consider specifying any auction requirement if it wishes to incorporate market economics more fully into compliance decisions.
- EU-ETS analysis suggests the most important variables in determining Phase 1 allowance price changes were oil and natural gas price changes; this apparent linkage raises possible market manipulation issues, particularly with the inclusion of financial instruments such as options and futures contracts. The EU will examine the matter in preparation for Phase 3. Congress may consider whether the government needs enhanced regulatory and oversight authority over such instruments.

OVERVIEW[1]

Climate change is generally viewed as a global issue, but proposed responses typically require action at the national level. With the 1997 Kyoto Protocol now in force and setting emissions objectives for 2008-2012, countries that ratified the protocol are implementing strategies to begin reducing their emissions of greenhouse gases.[2] In particular, the European Union (EU) has decided to use an emissions trading scheme (called a "cap-and-trade" program), along with other market-oriented mechanisms permitted under the Protocol, to help it achieve compliance at least cost.[3] The decision to use emission trading to implement the Kyoto Protocol is at least partly based on the successful emissions trading program used by the United States to implement its sulfur dioxide (acid rain) control program contained in Title IV of the 1990 Clean Act Amendments.[4]

The EU's Emissions Trading System (ETS) covers more than 10,000 energy-intensive facilities across the 27 EU Member countries, including oil refineries, powerplants over 20 megawatts (MW) in capacity, coke ovens, and iron and steel plants, along with cement, glass, lime, brick, ceramics, and pulp and paper installations. In addition, aviation is currently being phased into the ETS. These covered entities emit about 40%-45% of the EU's total greenhouse gas emissions, and almost two-thirds of them are combustion installations. The trading program does not cover either carbon dioxide (CO_2) emissions from the transportation sector (except aviation), which account for about 25% of the EU's total greenhouse gas emissions, or emissions of non-CO_2 greenhouse gases, which account for about 20% of the EU's total greenhouse gas emissions. A Phase 1 trading period ran between January 1, 2005, and December 31, 2007.[5] A Phase 2 trading period began January 1, 2008, covering the period of the Kyoto Protocol, and a Phase 3 has been finalized to begin in 2013.[6]

Under the Kyoto Protocol, the then-existing 15 nations of the EU agreed to reduce their aggregate annual average emissions for 2008-2012 by 8% from the Protocol's baseline level (mostly 1990 levels) under a collective arrangement called a "bubble." In light of the Kyoto Protocol targets, the EU adopted a directive establishing the EU-ETS that entered into force October 13, 2003.[7]

One objective of the second phase of the ETS is to achieve 3.3 percentage points of the 8.0% reduction required by the EU-15 under the Protocol.[8]

The importance of emissions trading was elevated by the accession of 12 additional central and eastern European countries to EU membership from

May 2004 through January 2007. For the new EU-27, the overall ETS emissions cap is set at 2.08 billion metric tons of carbon dioxide (CO_2) annually for the Kyoto compliance period (2008-2012).

The second phase Kyoto compliance stage of the ETS is built on the experience the EU gained from its preliminary Phase 1. The European Commission (EC) believes that the Phase 1 "learning by doing" exercise prepared the community for the difficult task of achieving the reduction requirements of the Kyoto Protocol. Several positive results from the Phase 1 experience assisted the ETS in making the Phase 2 process run smoothly, at least so far. First, Phase 1 established much of the critical infrastructure necessary for a functional emission market, including emissions monitoring, registries, and inventories. Much of the publicized difficulty the ETS experienced early in the first phase can be traced to inadequate emissions data infrastructure.[9] Phase 1 significantly improved those critical elements in preparation for Phase 2 implementation.

Second, the ETS helped jump-start the project-based mechanisms—Clean Development Mechanism (CDM) and Joint Implementation (JI)—created under the Kyoto Protocol.[10] As stated by Ellerman and Buchner:

> The access to external credits provided by the Linking Directive has had an invigorating effect on the CDM and more generally on CO_2 reduction projects in developing countries, especially in China and India, the two major countries that will eventually have to become part of a global climate regime if there is to be one.[11]

Third, according to the EC, a key result of Phase 1 was its effect on corporate behavior. An EC survey of stakeholders indicated that many participants are incorporating the value of allowances in making decisions, particularly in the electric utility sector, where 70% of firms stated they were pricing the value of allowances into their daily operations, and 87% into future marginal pricing decisions. All industries stated that it was a factor in long-term decision-making.[12]

However, several issues that arose during the first phase remained contentious as the ETS implemented Phase 2, including allocation (including use of auctions and reliance on model projections), new entrant reserves, and others. In addition, the expansion of the EU and the implementation of the linking directives created new issues to which Phase 2 has had to respond.[13] Based on lessons learned in Phase 1 and Phase 2, the EU has taken a substantially different approach to these issues in Phase 3 that is discussed later.

RESULTS FROM PHASE 1 AND 2

It is unclear to what degree the first phase of the ETS achieved real emissions reductions. Emissions are dynamic over time; a product of a country's population, economic activity, and greenhouse gas intensity.[14] To capture these dynamics, each Member State of the EU developed an emissions baseline from models that project future trends in the country's emissions based on these and other factors, such as anticipated energy and greenhouse gas policies.[15] During the first phase, the emissions goal was to put the EU on the path to Kyoto compliance—not actually comply with the Protocol (which wasn't necessary until the 2008-2012 time period). Thus, countries developed "business as usual" baselines based on projected growth in emissions. Such a projected baseline suffers from two sources of uncertainty: data uncertainties, and forecasting uncertainties. On data, Phase 1 suffered from uncertainties with respect to data collection and coverage, in monitoring methods for historic data, and data verification. On projecting future emissions, Phase 1 faced uncertainties with respect to economic or sector-based growth rates. Fueled in many cases by over-optimistic economic growth assumptions, these uncertainties increased the probability of inflated business as usual baselines.[16]

The combination of these factors and modest reduction requirements resulted in the emissions allocations for the 2005-2007 trading period being higher than actual 2005 emissions.[17] This result raised questions about how much reductions achieved during Phase 1 were real as opposed to being merely paper artifacts. On the positive side, verified emissions in 2005 were 3.4% below the estimated 2005 baseline used during the allocation process. In addition, the allowance prices for 2005 stayed persistently high, suggesting some abatement was occurring and raising questions of "windfall" profits. As stated by Ellerman and Buchner:

> First, and most importantly, the persistently high price for EUAs [EU emissions allowances] in a market characterized by sufficient liquidity and sophisticated players must be considered as creating a presumption of abatement. It would be startling if power companies did not incorporate EUA prices into dispatch decisions that would have shifted generation to less emitting plants. There is plenty of anecdotal evidence that this was the case,

and the prominent charges of windfall profits assume that the opportunity cost of freely allocated allowances was being passed on (without noting the implications for abatement). Similarly, it would be surprising if there were no changes in production processes that could be made by the operators of industrial plants.[18]

However, EU emissions allowances (EUAs) during Phase 1 did not maintain value. Phase 1 EUAs were basically worthless during the final six months of 2007. This decline in EUA prices at least partially reflected the general non-transferability of Phase 1 EUAs to Phase 2. Only Poland and France included limited banking in their Phase 1 implementation plans (called National Allocation Plans (NAPs)). The EC further restricted use of Phase 1 EUAs in Phase 2 with a ruling in November 2006.[19] As a result, excess Phase 1 EUAs were worthless at the end of 2007.[20]

One consequence of the non-transferability of Phase 1 EUAs is that prices for Phase 2 EUAs remained relatively firm until the 2008-2009 recession reduced demand, as indicated by Figure 1. Scarcity is critical for the proper functioning of an allowance market. As further indicated by Figure 1, during 2009, the market firmed up at a much lower level as participants assessed the impact of the recession on the demand for EUAs. This is a different response than the market had during Phase 1, and may reflect Phase 2 improvements in the system. In particular, the more predictable 2009 response may reflect the ability of the EC to certify Phase 2 NAPs using more verifiable baseline data than were available for Phase 1.[21] A major reason the EC rejected *ex post* adjustments[22] was fear that such adjustments would have a disruptive effect on the marketplace.[23] Phase 1 did not firmly establish this foundation of markets;[24] based on the Phase 2 EUA future's market, further market development appears to be occurring, although, like most commodity markets, it remains somewhat volatile at times.

While the environmental performance of Phase 1 may be disputed, the need for additional reductions to achieve Kyoto is not. For 2008, the EU-15 is estimated to be 6.2% below its base-year emissions, compared with an 8% five-year average reduction commitment under the Kyoto Protocol. As indicated by Figure 2, this represents a continuation of reductions by EU-15 over the past five years. However, as indicated by the pink line, the European Environment Agency (EEA) projects that the EU-15 existing measures are insufficient to reduce EU-15 emissions to their Kyoto requirements (represented by the purple line), resulting in a projected 6.9% reduction from baseline levels. To achieve the Kyoto target the EU projects further actions

reductions by EU-15 countries (represented by the green line in Figure 2), resulting in an overall reduction of 8.5% compared with baseline levels.

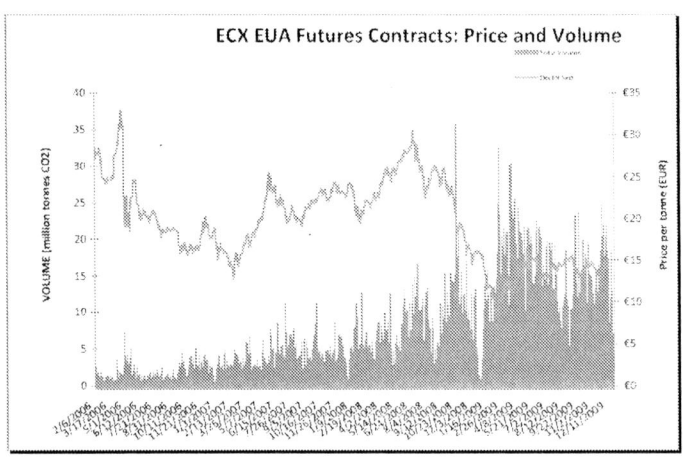

Source: ECX Exchange.
Note: Dec09 Sett: Future contracts with a settlement date of December, 2009.

Figure 1. ECX CFI Futures Contracts: Price and Volume

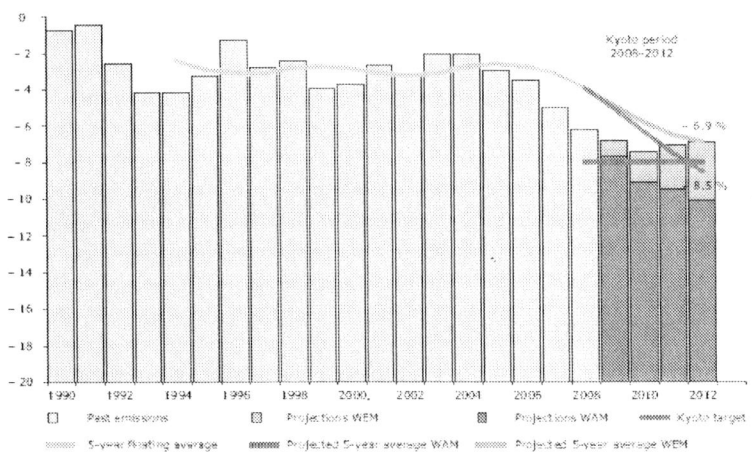

Source: European Environmental Agency, *Greenhouse Gas Emissions Trends and Projects in Europe 2000*, (2009) p. 10.
Note: WEM: with existing measures (measures implemented or adopted). WAM: with additional measures (planned measures).

Figure 2. EU-15 Greenhouse Gas Emissions and Projections for the Kyoto Period: 1990-2012

In addition to domestic emission reductions, the EU has also projected additional reduction credits received by activities permitted under the Kyoto Protocol: (1) purchase of project-based credits by ETS participants and EU governments (e.g., Joint Implementation (JI) and Clean Development Mechanism (CDM) projects); and, (2) the use of carbon sinks. As indicated in Figure 3, these activities provide a credit on the EU-15 baseline of 4.6 percentage points. Thus, if the EU-15 maintains its current path, it would exceed its Kyoto commitment by about 3.5 percentage points (6.9% minus 3.4%). If its planned measures result in the projected 8.5% reduction below baseline levels, the overachievement of its Kyoto commitment would be 5.1 percentage points (8.5% minus 3.4%).[25]

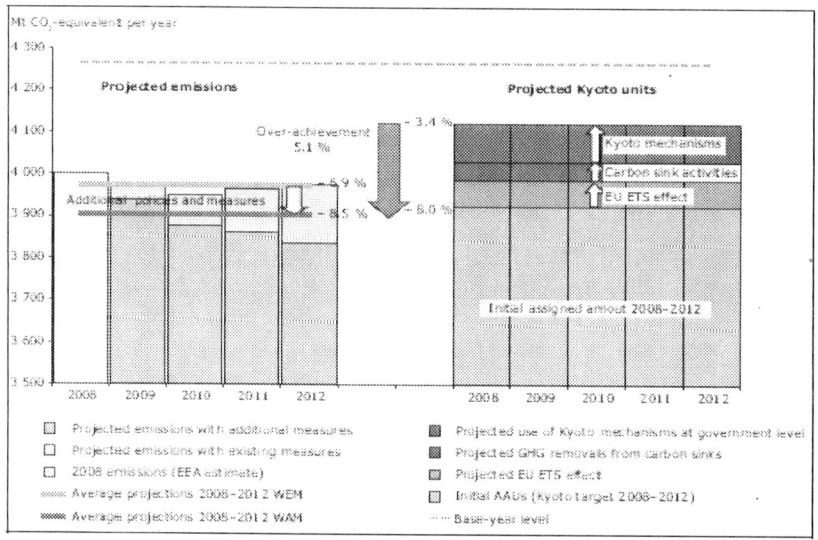

Source: European Environmental Agency, *Greenhouse Gas Emissions Trends and Projects in Europe 2000*, (2009) p. 11.

Notes: The left section shows the projected emissions considering only domestic measures (existing and additional) and is showing them as average 2008-2012 emissions (lines) and annual emissions (bars). The right section shows the projected amount of Kyoto credits accumulated by the end of the commitment period, including the initial EC assigned amount under the Protocol, the purchase of Kyoto project credits by EU ETS participants and EU governments, and carbon sink activities.

Figure 3. Summary of EU-15 Emissions Projection Compared to Projected Kyoto Protocol Credits

The EU-27 as a whole does not have an emissions target comparable to the EU-15 bubble. By 2010, EU-27 emissions are projected at 9.6% below Kyoto baseline levels assuming current policies. This reduction is projected at 11.3% if additional measures are included. Currently, 24 of the 25 countries with reduction requirements are projected to meet their commitments under the Kyoto Protocol.[26] Only Austria is not projected to meet its requirements even with additional planned measures and the use of Kyoto mechanisms.[27]

PHASE 3

The European Union is committed to achieving a 20% reduction in greenhouse gas emissions by 2020 from 1990 levels (or more depending on the actions of other countries). A strategic component of the effort to achieve this target is a revised ETS that will achieve a 21% reduction from covered entities from 2005 levels. Table 1 indicates the proposed EU-wide ETS cap for the next Phase of EU greenhouse gas program (Phase 3) assuming no further international commitments (the final 2013 cap figure is required by June 30, 2010).

Table 1. Proposed Annual ETS Cap Figures for Phase 3

Year	Billion metric tons of CO_2e
Annual limit for Kyoto compliance period (2008-2012)	2.083
2013	1.974
2014	1.937
2015	1.901
2016	1.865
2017	1.829
2018	1.792
2019	1.756
2020	1.720

As indicated, the EC envisions a linear reduction in the ETS cap to match the reduction target under the overall 20% reduction program. These numbers will change as individual countries decide to include more facilities under the ETS and as the EC expands ETS coverage to include other sectors and non-CO_2 greenhouse gases.

For Phase 3, the EU is re-shaping the ETS to improve its efficiency and eliminate some of the problems identified during Phase 1 and 2.[28] For Phase 2, the improved emissions inventories resulting from Phase 1 allowed the EC to harmonize the types of installations covered by the ETS across the various Member States.[29] In addition, the EC imposed a uniform rule on the Member States preventing the use of *ex-post* adjustments. However, Phase 2 made little advancement in harmonizing individual countries' allocations schemes.[30] As with Phase 1, countries continue to differ widely on several key points.

The critical structural change the EU would make in Phase 3 is eliminating National Allocation Plans (NAPs), and replacing them with EU-wide rules with respect to allowance availability, allocations, and auctions. NAPs are central to the EU's effort to achieve its Kyoto obligations under Phase 2. Each Member of the EU submitted a NAP that lays out its allocation scheme under the ETS, including individual allocations to each affected unit. These NAPs were assessed by the EC to determine compliance with 12 criteria delineated in an annex to the emissions trading directive.[31] Criteria included requirements that the emissions caps and other measures proposed by the Member State were sufficient to put it on the path toward its Kyoto target, protections against discrimination between companies and sectors, and delineation of intended use of CDM and JI credits for compliance, along with provisions for new entrants, clean technology, and early reduction credits. For the second trading period, the NAP had to guarantee Kyoto compliance.

This NAP structure will be replaced under Phase 3. There would be one EU-wide cap instead of the 27 national caps under Phase 1 and 2. Allowances would be allocated under EU-wide, fully harmonized rules, including those governing: (1) auctions, (2) transitional free allocations for greenhouse gas intensive, trade-exposed industries, (3) new entrants, and (4) coverage. The EC proposed a Directive to alter the EU-ETS structure for Phase 3 in January, 2008,[32] and the Directive was amended and adopted by the European Parliament (EP) and of the Council of the European Union in April 2009.[33]

Auctions

Under Phases 1 and 2, allowances generally were and are allocated free to participating entities under the ETS. During Phase 1, The EU-ETS Directive allowed countries to auction up to 5% of allowance allocations, rising to 10% under Phase 2.[34] Under Phase 1, only 4 of 25 countries used auctions at all, and only Denmark auctioned the full 5%. The political difficulty in instituting significant auctioning into ETS allowance allocations is the almost universal agreement by covered entities in favor of free allocation of allowances and opposition to auctions.[35] Free allocation of allowances represents a one-time transfer of wealth to the entities receiving them from the government issuing them.[36] The resulting transfer of wealth has been described by several analysts as "windfall profits."[37] As summarized by Ellerman and Buchner: "Allocation in the EU ETS provides one more example that, notwithstanding the advice of economists, the free allocation of allowances is not to be easily set aside."[38]

Despite concerns about windfall profits and economic distortions resulting from the free allocation of allowances, there was little change in basic allocation philosophy for Phase 2. No country proposed auctioning the maximum percentage of allowances allowed (10%). Most do not include auctions at all.[39] The unwillingness of governments to employ auctions as an allocating mechanism revolve around equity considerations, including: (1) inability of some covered entities to pass through cost because of regulation or exposure to international competition; (2) potential drag on a sector's economic performance from the up-front cost of auctioned allowances; and (3) the potential that government will not recycle revenues to alleviate compliance costs, international competitiveness impacts, or other equity concerns, resulting in the auction costs being the same as a tax.[40]

This opposition is mostly overcome for Phase 3 through an EU-wide set of harmonized rules for allowance allocations and auctions. Under Phase 3, the Directive states:

> Auctioning should ... be the basic principle for allocation, as it is the simplest, and generally considered to be the most economic efficient system. This should also eliminate windfall profits and put new entrants and economies growing faster than average on the same competitive footing as existing installations. (paragraph 15)

After nine eastern European Member States threatened to veto an initial proposal to auction 100% of all allowances, the EU compromised to provide for some free allocation of allowances during Phase 3 that will begin in 2013.[41] Most covered industries will be eligible for some free allocation of allowances to cover direct emissions under the Phase 3 agreement. The introduction of auction would be differentiated by sector. In general, for the power sector, full auctioning will begin in 2013. For electric powerplants, most will receive no free allocation of allowances during Phase 3. However, in a concession to certain eastern European Member States, an optional and temporary derogation from the no-free-allocation requirement for powerplants is provided to countries that meet specific energy and economic criteria. Under the optional allocation scheme, the Member State can allocate allowances equal to 70% of the powerplant's Phase 1 emissions free; this allocation declines to zero in 2020.

The auction schedule for most other covered entities is more gradual with 80% of a sector's allocation provided free in 2013, declining linearly to 30% by 2020, and zero by 2027. As stated in the final Directive:

> For other sectors covered by the Community scheme, a transitional system should be foreseen for which free allocation in 2013 would be 80% of the amount that corresponded to the percentage of the overall Community-wide emissions throughout the period 2005 to 2007 that those installations emitted as a proportion of the annual Community-wide total quantity of allowances. Thereafter, the free allocation should decrease each year by equal amounts resulting in 30% free allocation in 2020, with a view to reaching no free allocation in 2027. (paragraph 21)

Because of concern that stringent EU carbon policies may encourage production and related greenhouse gas emissions to shift to countries without carbon policies (i.e., carbon leakage), exceptions to this phase-out of free allowances will be made in sectors where carbon leakage may occur, as discussed later.

Distribution of allowances to be auctioned by the Member States will be determined by a three-part formula (Article 10(2)). Eighty-eight percent of the allowances to be auctioned by each Member State is distributed to States according to their historic emissions under Phase 1 of the EU-ETS. Ten percent of the total is distributed to States mostly based in their comparative GDP per capita within the EU (Annex IIa). Two percent of the total is distributed to nine former eastern-bloc countries based on the substantial greenhouse gas reductions they have already achieved (Annex IIb). Auctions will be conducted at the Member State level (cooperative auctions between States are also allowed) and must be open to any potential buyer. The EC is directed to develop the appropriate rules for coordinated auctions by June 30, 2010.

Beyond the allocation of allowances, the EU Directive also provides guidelines for the allocation of revenues from allowance auctions. The Directive states that at least 50% of the proceeds should be used to fund a variety of climate change related activities, including emission reductions, adaptation activities, renewable energy, carbon capture and storage (CCS), the Global Energy Efficiency and Renewable Energy Fund, and assisting developing countries to avoid deforestation and increase afforestation and reforestation (Article 10(3)).

New Entrant Reserves

Unlike previous cap-and-trade programs, the EU-ETS includes provisions for allocating free allowances to new entrants to the system.[42] The reasoning behind this decision is based on equity: (1) it isn't fair to allocate allowances free to existing entities while requiring new entrants to purchase them, and (2) the EU doesn't want to put Member States at a disadvantage in competing for new investments.[43] These equity concerns trumped concerns about economic efficiency.

As is the case for existing entities, the free allocation of allowances to new entrants is a subsidy. Under Phase 1 and Phase 2, the size and distribution of this subsidy is left to the individual Member States. For Phase 1, the reserve varied widely from the average of 3% of total allowances: Poland set aside only 0.4% of its allocation for new entrants while Malta set aside 26%. For Phase 2, the spread continues with Poland reserving 3.2% of its allowances for new entrants in contrast to 45% reserved by Latvia.[44]

The decision to employ a new entrant reserve adds complexity to Member States' allocation plans and influences the investment decisions of covered entities. Rules had to be promulgated with respect to the reserve's size, manner in which the allowances are dispensed, and how to proceed if the demand either exceeds the supply, or vice versa. As indicated, countries did not harmonize new entrant reserve rules with respect to size during Phase 1 or 2. Likewise, there is no standardization on dispensing allowances and replenishing the reserve: first-come, first-serve with no replenishment is one approach used, but a variety of procedures have been developed both to dispense allowances and to replenish the reserve if supply is inadequate. Member States also have different formulas for determining how many allowances a new entrant should receive. Member States claim to use a form of "benchmarking" to determine allowance allocations—an approach based on a standard of "best practices" or "best technology" that is applied to the new entrant's anticipated production or capacity. However, the definitions and application of the benchmarks used by the Member States are not uniform.

This will change under Phase 3. Under Phase 3, the Directive sets an EU-wide cap of 5% of the total allowance cap for a new entrant reserve, and requires the harmonization of allocation rules. The EC is to adopt a harmonized rule for applying a new entrant definition contained in the Directive by December 31, 2010; the Directive expressly excludes any new electricity production from being defined as a new entrant. The EC is also to determine EU-wide benchmarks for the allocation of all free allowances. The Directive states that the starting point for setting those benchmarks shall be the average performance of the 10% most efficient installations in a sector or subsector in the EU in the years 2007-2008 (Article 10a(2)).

In an attempt to stimulate development of CCS, the Directive also provides that up to 300 million allowances in the new entrants' reserve shall be available through 2015 for aiding construction and operation of up to 12 demonstration projects. No one project can receive more than 15% of the allowances allocated for this purpose (Article 10a(8)).

EC Phase 3 Decision on Eligible Industries[45]

Most studies of the competitiveness impacts of the ETS during Phase 1 have found no impact. The International Energy Agency (IEA) cites several reasons for this situation:

> Experience to date with the EU-ETS does not reveal leakage for the sectors concerned—analysis of steel, cement, aluminum and refineries sectors reveals that no significant changes in trade flows and production patterns were evident during the first phase (2005-2007) of the EU-ETS. This is mostly due to the free allocation of allowances, sometimes in generous quantities, and to the still functioning long-term electricity contracts, which softened the blow of rising electricity prices. Further, the general boom in prices for most traded products subject to carbon costs—whether direct or indirect—has blurred any effects of the latter. Finally, the relatively short time span of these policies does not allow observation of the full potential effects on industry via changes in investment location decisions.[46]

This conclusion is echoed by Carbon Trust, which states that currently, free allocation of emissions allowances offset almost all of the additional costs of the ETS; and that conclusion is echoed by The Climate Group for The German Marshall Fund, which states that companies surveyed found it difficult to quantify effects on their bottom line in the first phase, or found no effect at all.[47]

For energy-intensive, trade-exposed industries, Phase 3 has provisions to provide assistance to eligible installations to address the direct and indirect impact of emissions control costs. With respect to direct emissions costs, the EC published a list of installations exposed to a significant risk of carbon leakage on December 24, 2009, as required under the Directive.[48] The list is identical to the draft list released in September 2009.[49] The decision lists 164 industrial sectors and subsectors deemed to be exposed sectors under the appropriate European Parliament and Council directives. Eligible installations will receive allowances sufficient to cover 100% of their direct emissions, provided they are using the most efficient technology available. This 100% allocation contrasts with the 80% distribution of free allowances to non-carbon leakage exposed industries in 2013. Reflecting the fluid nature of the competitive situation and international negotiations, the EC is to review its decision June 30, 2010, and provide the European Parliament and Council with any appropriate proposals to respond to the situation.

Assistance for the impact of indirect emissions control costs on exposed industries from higher electricity prices would be determined by Member States. As stated by the Directive:

> Member States may deem it necessary to compensate temporarily certain installations which have been determined to be exposed to a significant risk of carbon leakage related to greenhouse gas emissions passed on in electricity prices for these costs. Such support should only be granted where it is necessary and proportionate and should ensure that the Community scheme incentives to save energy and to stimulate a shift in demand from grey to green electricity are maintained. (paragraph 27)

Flexibility Mechanisms and Price Volatility Control

The major flexibility mechanism developed under the EU-ETS has been the Clean Development Mechanism (CDM) and Joint Implementation (JI) credits permitted under the Kyoto Protocol; however, this development has proven a controversial process. A major part of the controversy has been the "supplementarity" requirement of the Kyoto Protocol to use its flexibility mechanisms. Supplementarity requires that developed countries, such as most EU countries, ensure that their use of JI/CDM credits is supplemental to their own domestic control efforts. In defining supplementarity for Phase 2, the EC used 10% of a country's allowance allocation as a rule of thumb in approving NAPs—with a greater limit possible based on a country's domestic efforts to reduce emissions. This process resulted in some significant reductions in some countries' proposed limits (e.g., Ireland, Poland, Spain), but some increase in others (e.g., Italy, Latvia, Lithuania). Although these reductions appear substantial in individual cases, most analysts agree that they do not represent a major barrier to the cost-effective use of JI/CDM. However, the EU-ETS does not accept credits from land use, land-use change and forestry (LULUCF) projects.

For Phase 3, the EU maintains its ban on using LULUCF credits within the ETS. However, it will permit up to 50% of the required reductions mandated under Phase 3 to be achieved through CDM or JI credits. For existing installations, this represents a total of 1.6 billion credits over the eight-year compliance period. Limits on use of Kyoto credits will be based on a facility's 2008-2012 allocation (for an existing facilities) or its verified emissions during Phase 3 (for a new entrant or sector). The EC estimates that the minimum amount of Kyoto credits an existing facility will be able to use to comply with Phase 3 will be 11% of its 2008-2012 allocation, while new entrants and sectors will be able to use a minimum of 4.5% of their verified emissions during 2013-2020 (article 11a(8)). The precise percentages will be determined later.

Another flexibility mechanism, banking, is extended by the Directive from Phase 2 to Phase 3 in order to prevent a Phase 1 style collapse of allowance prices when the ETS transitions into Phase 3. In addition, the EU hopes that extending the trading period from five years to eight years, along with the steady, linear emissions reduction schedule, will increase certainty and stability in the allowance markets.

Phase 3 will introduce two other mechanisms designed to address price volatility. First, the EC is required under the Directive to examine whether the market for emission allowances is sufficiently protected from insider dealing or market manipulation. If not, the EC is to present proposals to ensure such protection to the EP and the Council (article 12(a)).

Second, the Directive provides that if the allowance price is more than three times the preceding two-year average for more than six consecutive months and the price is not based on market fundamentals, one of two measures may be taken. The first would allow Member States to shift forward the auctioning of some of its auctionable allowances. The second would allow

Member States to auction up to 25% of the remaining allowances in the new entrants reserve (article 29(a)).

Expanding Coverage

Despite the EC's interest in expanding the ETS, its coverage in terms of industries included for Phase 2 is essentially the same as for Phase 1. The exception is for aviation. In December, 2006, the EC proposed bringing greenhouse gas emissions from civil aviation into the ETS in two phases.[50] As agreed to by the European Parliament in July 2008, all intra-EU and international flights will be included under the ETS beginning in 2012. Emissions would be capped at 97% of average 2004-2006 emissions with 85% of the allowances being allocated free to operators. The cap would be reduced to 95% in 2013. The cap and auctioning of allowances would be reviewed as a part of Phase 3 implementation.

Annex I of the Directive identifies three CO_2 emitting sectors for inclusion under the ETS: petrochemicals, ammonia, and aluminum. The ETS will also expand beyond CO_2 to include nitrous oxide (N_2O) emissions from nitric, adipic, and glyoxalic acid production, and perofluorocarbon (PFC) emissions from the aluminum sector. This would expand ETS covered emissions by 4.6% over Phase 2 allowance allocations, or about 100 million metric tons.[51] The harmonization and codification of eligibility criteria for combustion installations is expected to increase the coverage by a further 40-50 million metric tons.

To improve the cost-effectiveness of the ETS and reduce administrative costs, the Directive provides that small installations may be subject to other control regimes (such as carbon taxes) rather than included under the EU-ETS. Currently, the smallest 1,400 (10% of total installations covered) installations emit only 0.14% of total emissions covered. The Directive provides that Member States may opt to exclude installations that emit less than 25,000 metric tons annually from the EU-ETS (paragraph 11).

Summary and Considerations for U.S. Cap-and-Trade Proposals

The United States is not a party to the Kyoto Protocol and no legislative proposal before the Congress would impose as stringent or rapid an emission reduction regime on the United States as Kyoto would have. Likewise, U.S. proposals to reduce emissions through 2020 are not as stringent as that provided in the EU Directive. However, through five years of carbon emissions trading, the EU has gained valuable experience. This experience, along with the process of developing Phase 3, may provide some insight into current cap-and-trade design issues in the United States.

Emission Inventories and Target Setting

The ETS experience with market trading and target setting confirms once again the central importance of a credible emissions inventory to a functioning cap-and-trade program.[52] The lack of credible EU-wide data on emissions was a direct cause of the ETS Phase 1 allowance market collapse in 2006. Arguably, the most important result of Phase 1 was the development of a credible inventory on which to base future targets and allocations.

In the United States, Section 821 of the 1990 Clean Air Act Amendments requires electric generating facilities affected by the acid rain provisions of Title IV to monitor carbon dioxide in accordance with EPA regulations.[53] This provision was enacted for the stated purpose of establishing a national carbon dioxide monitoring system.[54] As promulgated by EPA, regulations permit owners and operators of affected facilities to monitor their carbon dioxide emissions through either continuous emission monitoring (CEM) or fuel analysis.[55] The CEM regulations for carbon dioxide are similar to those for the acid rain program's sulfur dioxide CEM regulations. Those choosing fuel analysis must calculate mass emissions on a daily, quarterly, and annual basis, based on amounts and types of fuel used. As suggested by the EU-ETS experience, expanding equivalent data requirements to all facilities covered under a cap-and-trade program would be the foundation for developing allocation systems, reduction targets, and enforcement provisions.

Coverage

Despite economic analysis to the contrary, the EU decided to restrict Phase 1 ETS coverage to six sectors that represented about 40%-45% of the EU's CO_2 emissions.[56] This restriction was estimated to raise the cost of complying with Kyoto from 6 billion euro annually to 6.9 billion euro (1999 euro) compared with a comprehensive trading program. A variety of practical, political, and scientific reasons were given by the EC for the decision.[57]

The experience of the ETS up to now suggests that adding new sectors to an existing trading program is a difficult process. As noted above, a stated goal of the EC is to expand the coverage of the ETS. However, the experience of Phase 1 did not result in the addition of any new sector until the last year of Phase 2 when aviation will be included. The EU will expand its coverage with Phase 3, but the ETS will still cover fewer sectors emitting greenhouse gases than provided under most U.S. proposals.

U.S. cap-and-trade proposals generally fall into one of two categories. Most bills are more comprehensive than the ETS, covering 80% to 100% of the country's greenhouse gas emissions. At a minimum, they include the electric utility, transportation, and industrial sectors; disagreement among the bills center on the agricultural sector and smaller commercial and residential sources. In some cases discretion is provided EPA to exempt sources if serious data, economic, or other considerations dictate such a resolution.

A second category of bills focuses on the electric utility industry, representing about 33% of U.S. greenhouse gases and therefore less comprehensive than the ETS. Sometimes including additional controls on non-greenhouse gas pollutants, such as mercury, these bills focus on the sources with the most experience with emission trading and the best emissions data. Other sources could be added as circumstances dictate.

As noted, the EU's experience with the ETS suggests that adding sectors to an emission trading scheme can be a slow and contentious process. If one believes that the electric utility sector is a cost-effective place to start addressing greenhouse gas emissions and that there is sufficient time to do the necessary groundwork to eventually add other sectors, then a phased-in approach may be reasonable. If one believes that the economy as a whole needs to begin adjusting to a carbon-constrained environment to meet long term goals, then a more comprehensive approach may be justified. The ETS experience suggests the process doesn't necessarily get any easier if you wait.

Allocation Schemes

Setting up a tradeable allowance system is a lot like setting up a new currency.[58] Allocating allowances is essentially allocating money with the marketplace determining the exchange rate. As noted above, the free allocation scheme used in the ETS has resulted in "windfall profits" being received by allowance recipients. As stated quite forcefully by Deutsche Bank Research:

> The most striking market outcome of emissions trading to date has been the power industry's windfall profits, which have sparked controversy. We are all familiar with the background: emissions allowances were handed out free of charge to those plant operators participating in the emissions trading scheme. Nevertheless, in particular the producers of electricity succeeded in marking up the market price of electricity to include the opportunity-cost value of the allowances. This is correct from an accounting point of view, since the allowances do have a value and could otherwise be sold. Moreover, emissions trading cannot work without price signals.[59]

The free allocation of allowances in Phase 1 and 2 of the ETS incorporates two other mechanisms that create perverse incentives and significant distortions in the emissions markets: new entrant reserves and closure policy. Combined with an uncoordinated and spotty benchmarking approach for both new and existing sources, the result is a greenhouse gas reduction scheme that is influenced as much or more by national policy than by the emissions marketplace.

The expansion of auctions for Phase 3 of the ETS could simplify allocations and permit market forces to influence compliance strategies more fully. Most countries did not employ auctions at all during Phase 1 and auctions continue to be limited under Phase 2. No country combined an auction with a reserve price to encourage development of new technology. The EC limited the amount of auctioned allowances to 10% in Phase 2: a limit no country chose to meet. Efforts to expand auctions met opposition from industry groups, but attracted support from environmental groups and economists. The Phase 3 increased use of auctioning through 2020 will represent a major development for the scheme.

Currently, all U.S. cap-and-trade proposals have some provisions for auctions, although the amount involved is sometimes left to EPA discretion. Most specify a schedule that provides increasing use of auctions from 2012 through the mid-2030s with a final target of 66%-100% of total allowances auctioned. Funds would be used for a variety of purposes, including programs to encourage new technologies. Some proposals include a reserve price on some auctions to create a price floor for new technology.

Like the situation in the ETS, most U.S. industry groups either oppose auctions outright or want them to be supplemental to a base free allocation. Given the experience with the ETS where the EC and individual governments have been unwilling or unable to move away from free allocation, the Congress, like the EU, may ultimately be asked to consider specifying any auction requirement if it wishes to incorporate market economics more fully into compliance decisions.

Flexibility and Price Volatility

Despite EU rhetoric during the Kyoto Protocol negotiations, it moved into Phase 2 without a significant restriction on the use of CDM and JI credits. This embracing of project credits will significantly increase the flexibility facilities have in meeting their reduction targets. In addition, Phase 2 includes the use of banking to increase flexibility across time by allowing banked allowances to be used in Phase 3. Each of these market mechanisms is projected to reduce both the EU's Kyoto compliance costs and allowance price volatility. These flexibility mechanisms will be extended into Phase 3 with modifications.

Unfortunately, Phase 1 experience with the ETS did not provide much useful information on the value of market mechanisms or financial instruments in reducing costs or price volatility. The combination of poor emissions inventories, non-use of project credits, and time-limited allowances with effectively no banking resulted in extreme price volatility in Spring 2006, and virtually worthless allowances by mid-2007. The real test for the mechanisms employed by the ETS to create a stable allowance market is Phase 2. Initial indications are that a mature market for allowances appears to be developing, although, like most commodities markets, the allowance market can still be volatile at times.

Phase 3 is introducing two new mechanisms in the ETS to further respond to volatility not based on market fundamentals. However, the actual effectiveness of these mechanisms is yet to be proven.

Like the ETS, U.S. cap-and-trade proposals would employ a combination of devices to create a stable allowance market and encourage flexible, cost-effective compliance strategies by participating entities. All include banking. All include use of offsets, although some would place substantial restrictions on their use. Some proposals have incorporated a "safety valve" that would effectively place a ceiling on allowance prices, while others would create a Carbon Market Efficiency Board to observe the allowance market and implement cost-relief measures if necessary. Finally, some incorporate strategic reserves auctions, similar in concept to the EU forward auctioning mechanism, to increase allowance supply without busting the emission cap. Some see this as a more flexible response with the potential for avoiding or mitigating the environmental impacts of a safety valve (i.e., increased emissions).

Additionally, concern has been expressed in the United States about the regulation of allowance markets and instruments. Based on experience with the ETS, the potential for speculation and manipulation could extend beyond the emission markets. Analysis of ETS allowance prices during Phase 1 suggests the most important variables in determining allowance price changes were oil and natural gas price changes.[60] This apparent linkage between allowance price changes and price changes in two commodities markets raises the possibility of market manipulation, particularly with the inclusion of financial instruments such as options and futures contracts. The concern is sufficient for the Directive to require the EC to examine the situation and the current protections against such activities. Congress may ultimately be asked to consider whether the Securities and Exchange Commission, Federal Energy Regulatory Commission, the Commodities Futures Trading Commission, or other body should have enhanced regulatory and oversight authority over such instruments.[61]

End Notes

[1] Readers unfamiliar with the workings of the European Union may want to read CRS Report RS21372, *The European Union: Questions and Answers*, by Kristin Archick and Derek E. Mix.

[2] Six gases are included under the Kyoto Protocol: carbon dioxide, methane, nitrous oxide, hydrofluorocarbons, perfluorocarbons, and sulfur hexafluoride. The United States has not ratified the Kyoto Protocol and, therefore, is not covered by its provisions. For more information on the Kyoto Protocol, see CRS Report RL33826, *Climate Change: The Kyoto Protocol, Bali "Action Plan," and International Actions*, by Jane A. Leggett.

[3] Norway, a non-EU country, also has instituted a CO_2 trading system which is currently linked with the EU-ETS. Various other countries and a state-sponsored regional initiative located in the northeastern United States involving several states are developing mandatory cap-and-trade system programs, but are not operating at the current time. For a review of these emerging programs, along with other voluntary efforts, see CRS Report RL33812, *Climate Change: Action by States to Address Greenhouse Gas Emissions*, by Jonathan L. Ramseur.

[4] P.L. 101-549, Title IV (November 15, 1990).

[5] For further background on the ETS, see CRS Report RL34150, *Climate Change and the EU Emissions Trading Scheme (ETS): Kyoto and Beyond*, by Larry Parker.

[6] More information, including relevant directives, on the EU-ETS is available on the European Union's website at http://europa.eu.int/scadplus/leg/en/lvb/l28012.htm.

[7] Directive 2003/87/EC of the European Parliament and of the Council of 13 October 2003 establishing a scheme for greenhouse gas emissions allowance trading within the Community and amending Council Directive 96/61/EC.

[8] Commission of the European Communities, *Communication from the Commission: Progress towards Achieving the Kyoto Objectives* (November 19, 2008). Other reductions are to be achieve through regulatory measures, such as a CO_2 emissions standard for automobiles.

[9] A. Denny Ellerman and Barbara K. Buchner, "The European Union Emissions Trading Scheme: Origins, Allocations, and Early Results," *Environmental Economics and Policy* (Winter 2007), pp. 69-70; and International Emissions Trading Association, "IETA Position Paper on EU ETS Marking Functioning," (no date), p. 3.

[10] For more on the effect of the ETS on Kyoto mechanisms, see A. Denny Ellerman and Barbara K. Buchner, "The European Union Emissions Trading Scheme: Origins, Allocations, and Early Results," *Environmental Economics and Policy* (Winter 2007), p. 84; and International Emissions Trading Association, "IETA Position Paper on EU ETS Market Functioning" (no date), p. 2. For more information on the Kyoto Protocol mechanisms, see CRS Report RL33826, *Climate Change: The Kyoto Protocol, Bali "Action Plan," and International Actions*, by Jane A. Leggett.

[11] A Denny Ellerman and Barbara K. Buchner, "The European Union Emissions Trading Scheme: Origins, Allocations, and Early Results," *Environmental Economics and Policy* (Winter 2007), p. 84.

[12] European Commission, Directorate General for Environment, *Review of EU Emissions Trading Scheme: Survey Highlights*, (November 2005), pp. 5-7.

[13] For a further discussion of Phase 2 implementation issues, see CRS Report RL34150, *Climate Change and the EU Emissions Trading Scheme (ETS): Kyoto and Beyond*, by Larry Parker.

[14] For more information, see CRS Report RL33970, *Greenhouse Gas Emission Drivers: Population, Economic Development and Growth, and Energy Use*, by John Blodgett and Larry Parker.

[15] On the role of modeling in the first phase, see A Denny Ellerman and Barbara K. Buchner, "The European Union Emissions Trading Scheme: Origins, Allocations, and Early Results," 1 *Environmental Economics and Policy* 1 (Winter 2007), pp. 72-73.

[16] Regina Betz and Misato Sato, "Emissions Trading: Lessons Learnt from the 1st Phase of the EU ETS and Prospects for the 2nd Phase," 6 *Climate Policy* (2006), p. 354.

[17] For a further discussion, see CRS Report RL33581, *Climate Change: The European Union's Emissions Trading System (EU-ETS)*, by Larry Parker.

[18] A Denny Ellerman and Barbara K. Buchner, "The European Union Emissions Trading Scheme: Origins, Allocations, and Early Results," 1 *Environmental Economics and Policy* 1 (Winter 2007), p. 83.

[19] European Commission, *Communication from the Commission to the Council and to the European Parliament on the assessment of national allocation plans for the allocation of greenhouse gas emission allowances in the second period of the EU Emissions Trading Scheme*, COM(2006) 725 final (November 29, 2006), p. 11.

[20] For a further discussion, see Joseph Kruger, Wallace E. Oates, and William A. Pizer, "Decentralization in the EU Emissions Trading Scheme and Lessons for Global Policy, 1 *Environmental Economics and Policy* 1 (Winter 2007), p. 126; and, Frank J. Convery and Luke Redmond, "Market and Price Development in the European Union Emissions Trading Scheme, 1 *Environmental Economics and Policy* 1 (Winter 2007), pp. 96-7, 107.

[21] International Emissions Trading Association, "IETA Position Paper on EU ETS Market Functioning," (no date), p. 2.

[22] Once the EC has approved a country's NAP, including the total number of allowances and the allocation to each covered entity, the allocations can not be re-visited. Attempts to include provisions permitting such post-approval adjustments to a facility's allocation have been uniformly rejected by the EC.

[23] European Commission, *Communication from the Commission to the Council and to the European Parliament on the assessment of national allocation plans for the allocation of greenhouse gas emission allowances in the second period of the EU Emissions Trading Scheme,* COM(2006) 725 final (November 29, 2006), p 8; and, A Denny Ellerman and Barbara K. Buchner, "The European Union Emissions Trading Scheme: Origins, Allocations, and Early Results," 1 *Environmental Economics and Policy* 1 (Winter 2007), p. 71.

[24] On the mixed record of the EU-ETS and the need for allowance scarcity to a functioning emissions market, see Eric Haymann, *EU Emission Trading: Allocation Battles Intensifying,* Deutsche Bank Research (March 6, 2007). For a generally positive view of ETS market development, see Frank J. Convery and Luke Redmond, "Market and Price Development in the European Union Emissions Trading Scheme, 1 *Environmental Economics and Policy* 1 (Winter 2007), pp. 97-106. For a more negative view, see Karsten Neuhoff, Federico Ferrario, Michael Grubb, Etienne Gabel, and Kim Keats, "Emissions Projections 2008-2012 Versus NAPs II," 6 *Climate Policy* 5 (2006), pp. 395-410.

[25] European Environment Agency, *Greenhouse Gas Emissions Trends and Projections in Europe 2009,* (2009) p. 11.

[26] Cyprus and Malta are not Annex 1 countries.

[27] European Environmental Agency, *Greenhouse Gas Emission Trends and Projections in Europe 2009* (2009), p. 12.

[28] European Commission, Directive 2009/29/EC *of the European Parliament and of the Council of 23 April 2009 amending Directive 2003/87/EC so as to improve and extend the greenhouse gas emission allowance trading system of the Community* (Brussels, April 23, 2009). Hereinafter referred to as the Directive.

[29] European Commission, *Limiting Global Change to 2 degrees Celsius: The Way Ahead for 2020 and Beyond* (Brussels, January 10, 2007), p. 23.

[30] Joachim Schleich, Regina Betz, and Karoline Rogge, *EU Emissions Trading—Better Job Second Time Around?* Fraunhofer Institute System and Innovation Research (February 2007), p. 23.

[31] Commission of the European Communities, Directive 2003/87/EC, available at http://eur-lex.europa.eu/LexUriServ/ LexUriServ.do?uri=OJ:L:2003:275:0032:0046:EN:PDF.

[32] European Commission, *Proposal for a Directive of the European Parliament and of the Council amending Directive 2003/87/EC so as to improve and extend the greenhouse gas emission allowance trading system of the Community* (Brussels, January 23, 2008).

[33] European Commission, *Directive 2009/29/EC of the European Parliament and of the Council amending Directive 2003/87/EC so as to improve and extend the greenhouse gas emission allowance trading system of the Community* (Brussels, April 23, 2009).

[34] For a further discussion of auctioning and the ETS, see Cameron Hepburn, *et. al.*, "Auctioning of EU ETS phase II allowances: how and why?" 6 *Climate Policy* (2006), pp. 137-160.

[35] A Denny Ellerman and Barbara K. Buchner, "The European Union Emissions Trading Scheme: Origins, Allocations, and Early Results," 1 *Environmental Economics and Policy* 1 (Winter 2007), p. 73.

[36] Joseph Kruger, Wallace E. Oates, and William A. Pizer, "Decentralization in the EU Emissions Trading Scheme and Lessons for Global Policy," 1 *Environmental Economics and Policy* 1 (Winter 2007), p. 114.

[37] E.g., Deutsche Bank Research, *EU Emission Trading: Allocation Battles Intensifying,* (March 6, 2007) pp. 2-3; and Regina Betz and Misato Sato, "Emissions Trading: Lessons Learnt from the 1st Phase of the EU ETS and Prospects for the 2nd Phase," 6 *Climate Policy* (2006), p. 353.

[38] A Denny Ellerman and Barbara K. Buchner, "The European Union Emissions Trading Scheme: Origins, Allocations, and Early Results," 1 *Environmental Economics and Policy* 1 (Winter 2007), p. 85.
[39] For more information, see CRS Report RL34150, *Climate Change and the EU Emissions Trading Scheme (ETS): Kyoto and Beyond*, by Larry Parker.
[40] Martina Priebe, *Distributional Effect of Carbon-allowance Trading* (Cambridge, January 12, 2007). Also, see Eurochambres, *Review of the EU Emission Trading System* (June 2007), p. 5.
[41] See *Position of the European Parliament adopted at the first reading on 17 December 2008 with a view to the adoption of Directive 2009/.../EC of the European Parliament and of the Council amending Directive 2003/87/EC so as to improve and extend the greenhouse gas emission allowance trading system of the Community* (December 17, 2008).
[42] For example, the U.S. acid rain program provides no allocation of allowances to new entrants; instead, an EPA sanctioned auction is held annually to ensure that allowances are available to new entrants. New entrants can also obtain allowances from existing sources willing to sell them, either directly, through the EPA auction, or via a broker.
[43] A Denny Ellerman and Barbara K. Buchner, "The European Union Emissions Trading Scheme: Origins, Allocations, and Early Results," 1 *Environmental Economics and Policy* 1 (Winter 2007), p. 75.
[44] Karoline Rogge, Joachim Schleich, and Regina Betz, *An Early Assessment of National Allocation Plans for Phase 2 of EU Emission Trading*, Fraunhofer Institute System and Innovation Research (January 2006).
[45] For more information on climate change and competitiveness issues, see CRS Report R40914, *Climate Change: EU and Proposed U.S. Approaches to Carbon Leakage and WTO Implications*, by Larry Parker and Jeanne J. Grimmett.
[46] Julia Reinaud, *Issues Behind Competitiveness and Carbon Leakage: Focus on Heavy Industry* (October 2008), p. 6.
[47] Carbon Trust, *EU ETS Impacts on Profitability and Trade* (January 2008), p. 4; and The Climate Group, *The Effects of EU Climate Legislation on Business Competitiveness; A Survey and Analysis* (September 2009), p. 8.
[48] European Commission, *Commission Decision of 24 December 2009 determining, pursuant to Directive 2003/87/EC of the European Parliament and of the Council, a list of sectors and subsectors which are deemed to be exposed to a significant risk of carbon leakage* (Brussels, 2009).
[49] European Commission, *Draft Commission Decision of 18 September 2009 determining, pursuant to Directive 2003/87/EC of the European Parliament and of the Council, a list of sectors and subsectors which are deemed to be exposed to a significant risk of carbon leakage* (Brussels, 2009).
[50] European Commission, *Proposal for a Directive of the European Parliament and of the Council amending Directive 2003/87/EC so as to include aviation activities in the scheme for greenhouse gas emission allowance trading within the Community* (Brussels, December 12, 2006).
[51] European Commission, *Proposal for a Directive of the European Parliament and of the Council amending Directive 2003/87/EC so as to improve and extend the greenhouse gas emission allowance trading system of the Community* (Brussels, January 23, 2008), p. 4.
[52] As stated by CRS in 1992: "For an economic incentive system to be effective, several preconditions are necessary. Perhaps the most important is data about the emissions being controlled. Such data are important to levy any tax, allocate any permits, and enforce any limit." CRS Issue Brief IB92125, *Global Climate: Proposed Economic Mechanisms for Reducing CO_2*, by Larry Parker (archived November 16, 1994), p. 9.
[53] Section 821, *1990 Clean Air Act Amendments* (P.L. 101-549, 42 USC 7651k).
[54] S.Rept. 101-952.

[55] See 40 CFR 75.13, along with appendix G (for CEMs specifications) and appendix F (for fuel analysis specifications).

[56] For more background, see CRS Report RL33581, *Climate Change: The European Union's Emissions Trading System (EU-ETS)*, by Larry Parker.

[57] Ibid., p 3.

[58] Unlike a carbon tax which uses the existing currency system to control emissions—be it euro or dollars.

[59] Deutsche Bank Research, *EU Emission Trading: Allocation Battles Intensifying* (March 6, 2007), p. 2.

[60] Maria Mansanet-Bataller, Angel Pardo, and Enric Valor, "CO_2 Prices, Energy and Weather," 28 *The Energy Journal* 3 (2007), pp. 73-92.

[61] For a discussion of regulation of allowances as a commodity and implications for a greenhouse gas emissions market, see CRS Report RL34488, *Regulating a Carbon Market: Issues Raised By the European Carbon and U.S. Sulfur Dioxide Allowance Markets*, by Mark Jickling and Larry Parker.

In: Countries and Climate Policies and Paths... ISBN: 978-1-61728-923-1
Editors: Thomas P. Parker © 2011 Nova Science Publishers, Inc.

Chapter 3

A. U.S.-CENTRIC CHRONOLOGY OF THE INTERNATIONAL CLIMATE CHANGE NEGOTIATIONS

Jane A. Leggett

SUMMARY

Under the 2007 "Bali Action Plan," countries around the globe sought to reach a "Copenhagen agreement" in December 2009 on effective, feasible, and fair actions beyond 2012 to address risks of climate change driven by human-related emissions of greenhouse gases (GHG). The Copenhagen conference was beset by strong differences among countries, however, and (beyond technical decisions) achieved only mandates to continue negotiating toward the next Conference of the Parties (COP) to be held in Mexico City in December 2010. The COP also "took note of" (not adopting) a "Copenhagen Accord," agreed among the United States and additional countries (notably including China), which reflects compromises on some key actions.

As background to the ongoing negotiations, this document provides a U.S.-centric chronology of the international policy deliberations to address climate change from 1979-2009. It begins before agreement on the United Nations Framework Convention on Climate Change (UNFCCC) in 1992, and proceeds through the Kyoto Protocol in 1997, the Marrakesh Accords of 2001, the Bali Action Plan of 2007, and the Copenhagen conference in 2009. The

Bali Action Plan mandated the Copenhagen negotiations on commitments for the period beyond 2012, when the first commitment period of the Kyoto Protocol ends. This chronology identifies selected external events and major multilateral meetings that have influenced the current legal and institutional arrangements, as well as contentious issues for further cooperation.

Negotiations underway since 2007 have run on two tracks: one under the Kyoto Protocol (which is subsidiary to the Convention), to extend commitments of developed, *Annex I*, Parties beyond 2012. This track excludes the United States, which is not a Party to the Kyoto Protocol and has said it will not join the Protocol. The second track proceeds directly under the Convention under the Bali Action Plan and focuses on five primary elements: a "shared vision" for reducing global GHG emissions by around 2050; mitigation of greenhouse gas emissions; adaptation to impacts of climate change; financial assistance to low income countries; and technology development and diffusion. Among the most difficult issues have been provisions for mutual assurance of compliance among Parties through *measurement, reporting, and verification* (MRV) of GHG emissions and removals, nationally appropriate mitigation actions, and financial and technical support from the wealthiest countries for adaptation, technology, and capacity-building. Some progress has been made on arrangements *to reduce emissions from deforestation and forest degradation* (REDD-plus). However, Parties did not reach consensus in Copenhagen on any of these elements, and the mandates for negotiation on the two tracks have been extended into 2010. The Copenhagen Accord may represent a supplemental or alternative track. Currently, the way forward remains unclear.

Many in the U.S. Congress are concerned with the goals and obligations that a treaty or other form of agreement might embody. A particular concern regards parity of actions and trade competitiveness effects among countries. For U.S. legislators, additional issues include the compatibility of any international agreement with U.S. domestic policies and laws; the adequacy of appropriations, fiscal measures, and programs to achieve any commitments under the agreement; and the desirable form of the agreement and related requirements, with a view toward potential Senate ratification of the agreement and federal legislation to assure that U.S. commitments are met.

OVERVIEW OF THE INTERNATIONAL CLIMATE CHANGE NEGOTIATIONS

Formal international negotiations were launched in December 1990 to address human-driven climate change. These negotiations on a Framework Convention on Climate Change marked the progress of decades of scientific research toward conclusions—with uncertainties—that have remained remarkably stable in the years since: greenhouse gas (GHG)[1] emissions from human-related activities are very likely causing the major portion of climate change observed in recent decades and, if these continue, could lead to potentially catastrophic impacts on human societies and their environment. Predicting the precise timing, magnitude and implications of changes remains subject to a variety of uncertainties; many questions may not be resolvable in a timeframe consistent with making effective and cost-effective decisions to address the risks of climate change. Only concerted global action can stabilize GHG concentrations, since emissions come from all countries. China has surpassed the United States as the leading emitter of GHG, although the United States historically has contributed more—almost one-fifth of the rise of GHG concentrations in the atmosphere. The greatest growth in GHG emissions is expected from countries, such as China, India and Brazil, that historically have contributed less, now emit much less per person, and have lower economic and governance capabilities to address the problem.

The core issues for negotiation in 1990 remain the same today:

- when and by how much to reduce greenhouse gas emissions globally in order to avoid *"dangerous anthropogenic interference with the climate system"*;[2]
- how to share *"common but differentiated responsibilities"* among countries taking into account *"historic contributions"* and *"respective capacities"* of different people—in particular, the acceptable degree of participation of developing countries;
- what mechanisms are best suited to assuring GHG reductions by all parties at the lowest cost, respecting national sovereignty and while supporting *"sustainable economic development"* and *"the eradication of poverty"*;
- how cooperatively to understand the risks and facilitate adaptation to climate changes, especially by those least able to cope on their own; and

- how to adapt international arrangements over time as science, social conditions, and capabilities evolve.

THE UNITED NATIONS FRAMEWORK CONVENTION ON CLIMATE CHANGE

The international negotiations launched in 1990 culminated in the 1992 adoption of the United Nations Framework Convention on Climate Change (UNFCCC) in Rio de Janeiro, Brazil. The United States was the fourth nation to ratify the UNFCCC, and the first among industrialized countries. As of November 2008, 192 governments are Parties to the UNFCCC. As a framework convention, this treaty provides the structure for collaboration and evolution of efforts over decades, as well as the first step in that collaboration. The UNFCCC does not, however, include measurable and enforceable objectives and commitments.[3] By the time the treaty entered into force and the Conference of the Parties (COP) met for the first time in 1995, the Parties agreed that achieving the objective of the UNFCCC would require new and stronger GHG commitments, though the Berlin Mandate deferred any new commitments for developing countries for future agreements. The resulting 1997 accord, the Kyoto Protocol, pledged to reduce the net GHG emissions[4] of industrialized country Parties (Annex I Parties) to 5.2% below 1990 levels in the period of 2008 to 2012. It also pledged to assess the adequacy of these commitments early in the new century.

THE KYOTO PROTOCOL

The United States signed the Kyoto Protocol in December 1997. However, opposition in the U.S. Congress was strong. In the "Byrd-Hagel" Resolution[5] in July 1997, the Senate expressed its opposition (95-0 vote) to the terms of the Berlin Mandate, by stating that the U.S. should not sign any treaty that does not include specific, scheduled commitments of non-Annex I Parties in the same compliance period as Annex I Parties, or that might seriously harm the U.S. economy. The Kyoto Protocol (KP) was not submitted to the Senate for ratification by President Clinton, nor by his successor, President George W. Bush. Newly elected President Bush announced in 2001 that the United States would oppose the agreement because it did not include GHG commitments by other large emitting (developing) countries and because of his conclusion that it would cause serious harm to the U.S. economy. As of November 1, 2008, 183 governments had become Parties to the Kyoto Protocol, with the United States and Kazakhstan[6] being the only industrialized countries to remain outside of the Kyoto Protocol. In KP Article 9, the Parties to The Kyoto Protocol agreed to begin a process no later than 2005 to consider commitments beyond 2012, when the first commitment period ends.

THE BALI ACTION PLAN AND KYOTO PROTOCOL TRACKS

In 2007, Parties agreed to establish two tracks for negotiation of further commitments of Parties. The first track was a mandate among the Kyoto Protocol Parties (not including the United States) to pursue an amendment to the Protocol on further commitments of Annex I Parties for period(s) beyond the year 2012. The first commitment period runs from 2008 through 2012.

The second track was established in December 2007, when the Conference of the Parties (COP) to the UNFCCC agreed to a "Bali Action Plan" to negotiate new GHG mitigation targets for Annex I Parties, "nationally appropriate mitigation actions" for non-Annex I Parties, and other commitments for the post-2012 period. The mandates specified that the products of negotiation should be ready by the end of 2009, for decision at the 15th meeting of the COP and the fifth meeting of the COP/MOP, in Copenhagen, Denmark. The form(s) of agreement were not clear, nor how the two negotiating tracks might converge.

The key items for the "Copenhagen" negotiations to address climate change beyond 2012 were:

- mitigation of climate change (primarily to reduce GHG emissions or to enhance removals of carbon by forests and other vegetation "sinks");
- adaptation to impacts of climate change;
- financial assistance to low income countries;
- technology development and transfer; and
- a shared vision for long-term goals and action.

In addition, provisions for "monitoring, reporting, and verification" (MRV) permeated the negotiations. Provisions to reduce GHG emissions from deforestation and forest degradation (REDD-plus) were also pursued under the Bali Action Plan.

Four meetings in 2008 and four in 2009 were scheduled, along with numerous inter-sessionals, regional group meetings, ministerials, and summits, in an ambitious attempt to reach an agreement of some kind by the Copenhagen meetings in December 2009. In Poznan, Poland, at the 14th COP, Parties decided to "shift into full negotiating mode" and that a first, full negotiating text should be available for a meeting in Bonn in June 2009. Under the UNFCCC, all bases for amending the Convention or a Protocol must be proposed at least six months before adoption.

THE COPENHAGEN SESSIONS AND THE "COPENHAGEN ACCORD"

It may take many months to evaluate the practical outcomes of the Copenhagen negotiations under the UNFCCC and the Kyoto Protocol. Despite the determination evidenced by participation by 193 delegations and 119 heads of state, strong disagreements on substance and process yielded results far below the (arguably unrealistic) expectations of many stakeholders. Formal decisions were largely technical. Both the COP and COP/MOP extended negotiating mandates into 2010. Progress was made on draft texts regarding some elements, such as pledges to commit to further GHG reductions by Annex I Parties, assistance for reducing emissions from deforestation and forest degradation, adaptation, and goals for financial assistance.

The sessions revealed distance among many countries' "bottom lines," leaving no space for consensus on major issues, such as the form and structure of agreements; obligations for GHG reductions and actions; whether commitments should be legally binding; and acceptable provisions for monitoring, reporting, and verification (MRV). Given the inability to reach consensus among the 193 delegations present, the United States, China, Brazil, India, and South Africa negotiated a "Copenhagen Accord" that bridges some difficult differences and identifies a common and differentiated path forward. While most UNFCCC Parties seemed willing to adopt the Copenhagen Accord, it was blocked by Bolivia, Cuba, Sudan, and Venezuela, arguing that the closed-door deal-making violated the procedures of the United Nations Charter. Tuvalu and some other nations rejected the agreement for not assuring, in their views, sufficiently deep GHG reductions. Consequently, the

COP only "took note" of the text, but did not adopt it. Hence, the Copenhagen Accord is a political outcome, not a legal agreement. Willing countries will be invited to join it. Nonetheless, President Obama was reported to have said "We should still drive toward something that is legally binding," a view held by most countries.

The Copenhagen Accord states a commitment ("shall") to "enhance our long-term cooperative action to combat climate change." In this regard, the Copenhagen Accord outlines a number of key points for action:

- **Long-term vision for GHG mitigation:** "Deep cuts" in global emissions are required "with a view to ... hold the increase in global temperature below 2 degrees C."
- **GHG mitigation by both Annex I and non-Annex I Parties:** Annex I Parties report GHG mitigation targets for 2020, and non-Annex I Parties report their mitigation actions, both before February 1, 2010, to be compiled in non-binding documentation. Least Developed Countries and small island developing states become a new mitigation grouping that may identify actions voluntarily and with financial support.
- **Transparent reporting and international review of Parties' mitigation while respecting national sovereignty:** Non-Annex I Parties must submit their National Communications bi-annually, and include reports on their domestic MRV of implementation of their mitigation actions, subject to *international consultations and analysis* that will ensure respect for national sovereignty. Mitigation actions (as well as technology, financing and capacity-building) supported by international finance will be subject to international MRV.
- **Immediate establishment of a mechanism including REDD-plus,**[7] to enable mobilization of international financing.
- **Goals for developed countries to mobilize finance for adaptation, mitigation, technology, and capacity-building:** Pledges of $30 billion during 2010-2012, and a goal of $100 billion annually by 2020 "in the context of meaningful mitigation actions and transparency on implementation." Funding will come from public and private, bilateral and multilateral, and alternative sources.
- **Establishment of the Copenhagen Green Climate Fund** under the Global Environment Facility, managed by the World Bank to support international financing.

- **Establishment of a Technology Mechanism** to "accelerate technology development and transfer" and to be "guided by a country-driven approach."
- **Assessment of the Copenhagen Accord,** to be completed by 2015, that would include consideration of strengthening the "long-term goal" of the Accord.

The Copenhagen Accord identifies a number of administrative decisions made to carry it out fully. These include development of guidelines for non-Annex I Parties' biannual National Communications; guidelines for "international consultations and analysis" of non-Annex I Parties' nationally appropriate mitigation actions; guidelines for MRV of nationally appropriate mitigation actions supported internationally; a mechanism to include REDD-plus; a mechanism of a High Level Panel to study potential sources of financing; the Copenhagen Green Climate Fund and its operational rules; a Technology Mechanism and its functions; and (after a few years) an assessment of implementation of the Accord.

A number of major proposals were notably not part of the final Copenhagen Accord:

- a target to avoid 1.5°C increase in global temperature (opposed by China);[8]
- that GHG emissions be cut by 2050 by 50% globally, and by Annex I countries by 80% (opposed by China; supported by a couple of developing countries);
- that there be a year by which global emissions would peak and then decline (opposed by China and other major developing country emitters);
- that non-Annex I countries reduce their emissions by 15-30% below business-as-usual projections by 2020 (opposed by China and other developing country emitters);
- specification of a baseyear (e.g., pre-industrial levels or 1990) for the aspirational target of avoiding global mean temperature increases of more than 2°C (3.6°F), (opposed by China);
- provisions that would make the Copenhagen Accord legally binding (opposed by China); and
- specification of specific amounts of funding to be pledged by individual Annex I Parties (opposed by the United States).

The process from Copenhagen to the next meeting of the COP, in Mexico City in December 2010, remains undefined. Two countries have offered to host related summits to facilitate progress. Bolivia has called an "alternative" meeting of indigenous peoples, social movements, environmentalists, scientists, and governments.

Fundamental disagreement remains on whether the outcome of further negotiations should be one or two agreements, or three if the Copenhagen Accord follow-up is included. The Copenhagen outcome leaves two separate negotiating mandates on the table, and no texts have been agreed on as the basis for further negotiations. COP-16 will be hosted by Mexico in December 2010. This meeting, which had been expected to set in motion the implementation of a Copenhagen agreement, may take on a very different function depending on processes and actions in 2010 that have yet to be defined. It also remains to be seen whether countries will be able and willing to move beyond the disarray and deadlocks witnessed in Copenhagen.

Congressional Interests in International Issues

International cooperation would be required to achieve the ranges of long-term targets for GHG mitigation and successful adaptation to climate change impacts. For U.S. legislators, assurance of actions by other major emitters is key to acceptability of U.S. mandates to abate emissions. Additional important issues include the compatibility of any international agreement with U.S. domestic policies and laws; the adequacy of appropriations and fiscal incentives to achieve any commitments under the agreement; and the desirable form of the agreement and any requirements for potential ratification and implementing legislation. Many Members of Congress are also attentive to questions of comparability of GHG actions among major trading partners, and especially to the potential for adverse competitiveness effects if some countries do not mandate GHG reductions.

U.S.-CENTRIC CHRONOLOGY OF INTERNATIONAL CLIMATE CHANGE NEGOTIATIONS, 1979-2009

1979   First World Climate Change Conference estimates that a doubling of carbon dioxide (CO_2) concentrations

	over pre-industrial levels would eventually lead to a 1.4-4.5°C increase in global mean temperature (GMT).
1987	In the Montreal Protocol, 57 governments agree to phase-out production of substances that deplete stratospheric ozone. Many of these substances, such as CFCs are also powerful and long-lasting greenhouse gases (GHG), implicated in climate change.
1985	Major scientific conference in Villach, Austria, reviews decades of observations and research, and calls for policy analysis and actions to slow the rate of GHG-induced climate change.
1988	Experts to the Toronto Conference on the Changing Atmosphere call for a reduction of global CO_2 emissions by 20% from 1988 levels by the year 2005.
November 1988	Governments establish the Intergovernmental Panel on Climate Change (IPCC) under the joint auspices of the UN World Meteorological Organization and the UN Environment Programme, to assess climate change research for governmental decision-making.
1990	Global CO_2 concentrations in the atmosphere are about 354 parts per million (ppm), compared to pre-industrial concentrations of about 280 ppm in 1750. Global CO_2 emissions are 21 billion tons annually, with 4/5 from industrialized countries (1/5 from the United States). Developing countries, home to 80% of the world's population, emit $1/5^{th}$ of global GHG emissions, not projected to reach 50% until around 2025.
1990	First Assessment Report of the IPCC concludes that human activities emit greenhouse gases (GHG) that have increased atmospheric concentrations; these may be causing observed increases in global mean temperature (GMT), and could drive future global warming. The human contribution could not be confirmed, however, for up to a decade.
1990	The United Nations General Assembly establishes the Intergovernmental Negotiating Committee for a Framework Convention on Climate Change.
June 1992	The United Nations Framework Convention on Climate Change (UNFCCC) opens for signature at the United

A. U.S.-Centric Chronology of the International Climate Change... 101

	Nations Conference on Environment and Development (UNCED) in Rio de Janeiro, Brazil. The treaty cites *common but differentiated responsibilities and respective capabilities* of all Parties, with an *objective* of *avoiding dangerous anthropogenic interference with the climate system*. It includes commitments of developed country "Annex I" Parties to establish national action plans with measures that aim (i.e., non-binding) to reduce GHG emissions to 1990 levels by the year 2000. Includes obligations for Parties listed in Annex II (including the United States) to provide technical and financial assistance, report GHG emissions, and additional obligations. The Global Environment Facility (GEF) is named the interim financial mechanism of the UNFCCC. Non-Annex I Parties have general obligations, including for GHG mitigation, adaptation planning, and reporting.
1 October 1992	The United States becomes the first industrialized nation to ratify the UNFCCC.
21 March 1994	Entry into Force of the UNFCCC, following ratification by 50 countries. (As of November 2008, 192 governments have ratified the UNFCCC.)
March-April 1995	In Berlin, Germany, the first meeting of the Conference of the Parties (COP-1) reviews the *adequacy of commitments* under UNFCCC Articles 4.2(a) and (b) and concludes they are inadequate. It therefore adopts the *Berlin Mandate*, initiating negotiations for the post-2000 period to strengthen the GHG commitments of Annex 1 Parties, but *no new commitments for non-Annex 1 Parties*. The COP also agrees to a Pilot Phase for Joint Implementation, and to establish two entities: the Subsidiary Body on Implementation (SBI) and the Subsidiary Body on Scientific and Technological Advice (SBSTA).
July 1997	The U.S. Senate passes (95-0) the *Byrd-Hagel Resolution*, that the United States should not enter into any international agreement that does not include obligations for developing countries in the same period, or that would seriously harm the U.S. economy.

December 1997	The *Kyoto Protocol* to the UNFCCC is adopted, signed by more than 150 countries. It sets a goal of reducing industrialized countries' GHG emissions to 5% below 1990 levels during the first commitment period of 2008-2012, and lists *assigned amounts* of allowable GHG emissions by Parties in Annex B. It provides for flexibility mechanisms, including trading of assigned amounts, Joint Implementation, and the Clean Development Mechanism. It outlines a compliance mechanism, and requires reporting by Parties. Many implementing rules remain to be negotiated, covering operations of the flexibility mechanisms, how to account for land-based carbon sequestration, the nature of the compliance regime, etc. The Protocol would enter into force when 55 countries, including at least 55% of 1990 GHG emissions, have submitted papers of ratification.
1998	The COP agrees to the *Buenos Aires Plan of Action*, with a deadline of 2000 to finalize rules to implement the Kyoto Protocol. The United States continues to press developing countries to take on voluntary commitments to reduce GHG emissions.
November 2000	In the Hague, Netherlands, the sixth COP discussions collapse, suspended without agreement on rules to implement the *flexibility mechanisms* in the Kyoto Protocol. Parties agree to resume talks at "COP-6bis" in July 2001.
January-May 2001	The IPCC releases its Third Assessment Report, concluding that global temperature and precipitation continue to increase, and effects can be observed in decreasing snow and ice extent, melting glaciers, altered seasonality, and other indicators of climate. The observed CO_2 concentration has not been exceeded during the past 420,000 years and likely not during the past 20 million years. Most of the observed warming over the last 50 years is likely due to the increased GHG concentrations, most of which results from fossil fuel use. Without concerted actions to abate GHG emissions, atmospheric CO_2 concentrations could rise to 540 to 970 ppm by

	2100—90 to 250% above the 280 ppm level in the year 1750. Associated global average temperature could rise over 1990 by 1.4° to 5.8°C (3.2°F to 14.4°F) by 2100; some regions would change more than others.
March 2001	President George W. Bush announces United States' opposition to the Kyoto Protocol, and becomes an Observer (not a Party) to deliberations concerning the Protocol.
July 2001	At COP-6bis, the United States participates for the first time as an observer, not a party to the Kyoto Protocol discussions. Decisions are made on use of the flexibility mechanisms (emissions trading, joint implementation and the Clean Development Mechanism), carbon sinks, emission penalties for non-compliance, and to establish three new financial mechanisms: the Special Climate Change Fund, the Least Developed Country Fund, and the Adaptation Fund.
December 2001	COP-7 adopts the *Marrakesh Accords*, establishing most rules and guidelines for the Kyoto Protocol to operate, especially for the three flexibility mechanisms: the Clean Development Mechanism, Joint Implementation, and Allowance Trading. To support adaptation in developing countries, agreements include: (1) replenishment of GEF to address needs of developing countries due to adverse effects of climate change or of response measures; (2) establishment of Special Climate Change Fund (SCCF) to support adaptation and technology transfer; (3) establishment of a Least Developed Country Fund (LDC Fund), with guidance on its operation; and (4) establishment of an Adaptation Fund under the Kyoto Protocol. The Parties also establish an LDC work program and the LDC Expert Group (LEG), funding for National Adaptation Programs of Action and additional implementation support. The United States participates for the first time as an Observer in deliberations related to the Kyoto Protocol.
November 2002	COP-8 issues a modest *Delhi Declaration on Climate Change and Sustainable Development.*

Summer 2003	Exceptional heat and air pollution in Western Europe are associated with more than 70,000 excess deaths. Scientific research indicated that global warming had at least doubled the chance of occurrence of the extreme heatwave.
30 October 2003	The first U.S. Senate vote on legislation to control GHG through a cap-and-emissions trading system, the McCain-Lieberman Climate Stewardship Act, fails (43-55), but gains more support than had been expected.
December 2003	COP-9 reaches several breakthrough decisions on credits for carbon absorption by forest sinks, as well as the Special Climate Change Fund (SCCF) and the Least Developed Countries Fund (LDC Fund).
November 2004	The Arctic Climate Impact Assessment concludes "Climate change, together with other stressors ... presents a range of challenges for human health, culture and well-being of Arctic residents ... as well as risks to Arctic species and ecosystems." Indigenous peoples link climate change impacts to human rights.
December 2004	COP-10 increases focus on adaptation and approves the Buenos Aires Programme of Work on Adaptation and Response Measures. Brazil and China submit their first National Communications to the UNFCCC.
1 January 2005	The European Union's Emissions Trading System (ETS) begins, permitting GHG allowance trading among 12 thousand companies.
16 February 2005	The Kyoto Protocol enters into force after Russia's ratification meets the requirement for ratification by Parties representing at least a 55% super-majority of CO_2 emissions (the requirement for at least 55 Parties to the UNFCCC having already been met).
2005	China announces ambitious energy efficiency and renewable energy policies.
25 June 2005	The U.S. Senate passes a Sense of the Senate Resolution (Amendment to H.R. 6) calling on Congress to enact "comprehensive and effective ... mandatory, market-based limits" to slow, stop, and reverse the growth of GHG emissions, at a rate and in a manner that would not "significantly harm" the U.S. economy.

27 July 2005	The United States announces the Asia-Pacific Partnership on Clean Development and Climate (APP), to cooperate on reducing the GHG intensity of their economies through voluntary technology exchanges. The APP includes the United States, Australia, Canada, China, India, Japan, and South Korea, and includes participation by the private sector.
November-December 2005	In Montreal, Canada, the first "Conference of the Parties serving as the Meeting of the Parties to the Kyoto Protocol" (CMP) meets. After the U.S. delegation walks out of the meeting, the COP agrees to two parallel tracks to consider actions in the post-2012 period, the Ad Hoc Working Group on Further Commitments for Annex I Parties under the Kyoto Protocol (AWG-KP), and another dialogue to be established under the UNFCCC.
6 June 2006	After a week of debate, the U.S. Senate rejects (38-60) the McCain-Lieberman proposal to establish a system of tradable allowances to reduce GHG emissions in the United States.
November 2006	In Nairobi, Kenya, COP-12 and CMP-2 reach agreements concerning the Adaptation Fund, the Nairobi Work Programme on Adaptation, and the Nairobi Framework on Capacity Building for the CDM.
10 January 2007	Commission of the European Union states a new policy of limiting global warming to 2° Celsius to reduce its GHG emissions unilaterally by 20% below 1990 levels by 2020, and to 30% below if other countries join in.
February-May 2007	The IPCC releases its Fourth Assessment Report, concluding that "warming of the climate system is unequivocal" and that "[m]ost of the observed increase in globally averaged temperatures since the mid-20th century is very likely due to the observed increase in anthropogenic GHG concentrations." By 2005, the global atmospheric concentration of CO_2 is 379 ppm, up 25 ppm since 1990, and up more than 35% over the pre-industrial level; the primary source of that increase is fossil fuel use and the second is land use change. While

	the United States adds about 18% of global GHG emissions, the emissions from China may have become the highest of any country.
April 2007	U.S. Supreme Court decides in *Massachusetts v. EPA* that GHG are air pollutants and that EPA must exercise the authority granted to it by the Clean Air Act to consider regulating these emissions.
May 2007	U.S. President Bush initiates the Major Economies Meetings (MEM) to negotiate a new post-2012 framework among a small group of countries, to develop a long-term global goal and "to complement ongoing UN activity."
31 August 2007	In Vienna, Parties to the Kyoto Protocol agree to consider a range of GHG reduction targets of 25% to 40% below 1990 levels for industrialized countries by 2020, though this range is resisted by Canada, Japan and Russia.
23 September 2007	At the first Major Economies Meeting (MEM), hosted by the United States, U.S. President George Bush pledges $2 billion over three years for a Clean Technology Fund (CTF) under the World Bank, expecting to raise $10 billion among donors to support concessional financing for energy projects in developing countries. Some environmental groups oppose inclusion of coal electricity in permitted project types.
December 2007	COP-13 agrees to the "Bali Action Plan"—establishes the Ad Hoc Working Group on Long-term Cooperative Action (AWG-LCA) with a mandate for Parties to the UNFCCC to negotiate toward new GHG mitigation actions and commitments in the post-2012 period and to reach agreement by the end of 2009 (at COP-14 meeting in Copenhagen, Denmark). The Bali Action Plan calls for "a shared vision for long-term cooperative action" and identifies 4 main elements: mitigation, adaptation, technology, and finance. Additional decisions place management of the Adaptation Fund under the World Bank, and initiate demonstrations and commitments to reduce deforestation.

A. U.S.-Centric Chronology of the International Climate Change... 107

15 May 2008	The U.S. Senate votes (55-40) that no new mandates on GHG should be enacted without effectively addressing imports from China, India and other nations without similar programs.
August 2008	In Accra, Ghana, exchange of views under the AWG-LCA continues on alternative approaches to "shared vision," mitigation, adaptation, technology and finance. Any question of differentiation among non-Annex I Parties continues to be contentious, with China and the G-77 maintaining solidarity. Some developing countries argue that the AWG-LCA and AWG-AP are not mandated to consider amendments to the UNFCCC or Kyoto Protocol, only implementation of them. Some delegations support worldwide sectoral approaches, which some developing countries argue would be inappropriate for them. Developing countries frequently call for new mechanisms for each issue, and oppose "conditionality" on financial and technology transfers (such as protection of intellectual property rights). The AWG-KP agree on a comprehensive "basket approach" to including multiple GHG in the second commitment period, and notes new groups of gases and new gases (e.g.,NF3) identified by the IPCC AR4. It notes that the Montreal Protocol phases out production of CFC and HCFC, but not their emissions. Analysis will proceed on various "spillover" effects of mitigation actions.
September 2008	Government of Japan proposes that all Parties adopt a "shared vision" of achieving at least 50% reduction of global GHG emissions by 2050. Global GHG emissions should peak in the next 10 to 20 years. It proposes criteria for entering additional countries into Annex I (i.e., to become countries with commitments), to create comparability of efforts for GHG targets among Annex I Parties, according to sectoral emissions, efficiencies, and reduction costs, and for new GHG commitments among three groups of developing countries.
December 2008	In Poznan, Poland, a high-level segment of COP-14 witnesses political statements on a "shared vision for long-term cooperative action," and agrees to intensify

	negotiations. Parties agree that a full negotiating text should be available by June 2009. Parties also resolve issues regarding the Adaptation Fund, though developing countries did not achieve commitments for additional adaptation monies. The Government of Mexico, among the first non-Annex I Parties to offer a GHG reduction commitment, announces a goal to halve GHG emissions from 2002 levels by 2050. Brazil pledges to cut deforestation by at least 50% by 2017.
1-12 June 2009	In Bonn, 30th sessions of the UNFCCC subsidiary bodies—SBSTA-30 and SBI-30; AWG-LCA-6 and AWG-KP- 8. Deliberation begins on a first negotiating text for a post-2012 agreement.
December 2009	COP-15 and COP/MOP-5 deliberate on multiple proposed texts without agreement, and decide to extend the negotiating mandates of AWG-LCA and AWG-KP through 2010. Key disagreements include whether the product should be two agreements (one being amendment of the Kyoto Protocol) or one merged text; whether obligations should be legally binding; and whether developing countries' mitigation actions and results should be measurable. COP-15 also "takes note of" the "Copenhagen Accord" negotiated among United States and roughly 30 countries outlining process to pledge (by February 1, 2010) national targets or actions to mitigate GHG emissions; $30 billion of financing from 2010-2012; and to seek $100 billion annually of a variety of types of financing by 2020.

End Notes

[1] "Greenhouse gases" are defined in the United Nations Framework Convention on Climate Change as "those gaseous constituents of the atmosphere, both natural and anthropogenic [human-driven] that absorb and re-emit infrared radiation." They may alter the composition of the atmosphere, changing the balance of radiation entering and leaving the Earth system, and consequently change the temperature or patterns of climate on Earth. The most important is water vapor, but it is believed not to be altered by human activities. Carbon dioxide (CO_2) is the most important human- related GHG, with about ¾ from fossil fuel use and about ¼ due to land use change and forestry. Other important gases listed under the

A. U.S.-Centric Chronology of the International Climate Change... 109

Kyoto Protocol are methane (CH_4), nitrous oxide (N_2O, hydrofluorocarbons (HFC), perfluorocarbons (PFC) and sulfur hexafluoride (SF_6). Additional greenhouse gases are partially controlled internationally under the *Montreal Protocol* of the Vienna Convention for the Protection of the Ozone Layer, including chlorofluorocarbons (CFC) and hydrochlorofluorocarbons (HCFC), etc., while others are emerging (e.g., nitrogen trifluoride (NF_3). Other radiatively important substances are significant but difficult to treat similarly, such as aerosols or tropospheric ozone.

[2] Terms used particularly in association with the international climate change negotiations are frequently highlighted in italics in this document, to alert the reader to their significance.

[3] The commitment by industrialized Parties to prepare national action plans aiming to reduce GHG emissions to 1990 levels is measurable, but no effective penalties or mechanisms were established to address any non-compliance with obligations.

[4] "Net" emissions are the gross emissions minus the removals of GHG from the atmosphere by "sinks" (sequestration), particularly by growing forests and other vegetation (or prevention of release of GHG by burning or decomposing vegetation).

[5] S.Res. 98.

[6] Kazakhstan is unusual in being considered an Annex I Party for the purposes of the Kyoto Protocol, but not for the purposes of the UNFCCC, once it ratifies the Kyoto Protocol [COP report FCCC/CP/2006/5].

[7] "REDD-plus" is Reducing Emissions from Deforestation and Forest Degradation plus enhancing carbon sequestration.

[8] Although many critics accuse the United States of a number of faults in the Copenhagen negotiations, a number have identified China as the sole obstacle to many points of potential agreement. One such account is Mark Lynas, "How Do I Know China Wrecked the Copenhagen Deal? I Was In the Room" The Guardian, London, December 22, 2009.

In: Countries and Climate Policies and Paths... ISBN: 978-1-61728-923-1
Editors: Thomas P. Parker © 2011 Nova Science Publishers, Inc.

Chapter 4

OPENING STATEMENT OF CHAIRMAN EDWARD J. MARKEY, THE SELECT COMMITTEE ON ENERGY INDEPENDENCE AND GLOBAL WARMING, HEARING ON "PREPARING FOR COPENHAGEN: HOW DEVELOPING COUNTRIES ARE FIGHTING CLIMATE CHANGE"

Over the last two years, the Select Committee has examined closely how the U.S. can fight climate change and improve our energy security. But we are not in this fight alone, and the progress that our country can make is deeply dependent on the progress that developing countries are making. That is the focus of today's hearing: to take an assessment based on the facts that exist in 2009 – not as they existed five or ten years ago – of steps taken by the key developing countries to address global warming.

This inquiry is important because Americans rightly want to know that they are not the only ones altering their policies to combat global warming. This inquiry is also important because many Members have rightly expressed concern about maintaining the competitiveness of critical industry sectors, and they want to know that other countries are joining the fight and requiring their industries to move away from business as usual.

A discussion on what developing countries are doing needs to be fuelled by current facts and not by old perceptions. One old perception is that China is unwilling to join the fight against climate change and is wedded to growth at any cost. A current reality is that China has already adopted an energy efficiency law that far exceeds anything on the law books of our country. Other examples abound of how developing countries are making progress.

I am not suggesting that the developing countries are doing everything that they can do, and they are certainly not doing everything that needs to be done. But as we undertake climate change legislation in our country we should understand the steps taken by key developing countries around the world.

China, India, Brazil, Mexico, South Africa are some of the biggest emitters in the developing world. Over the last years, all these countries have displayed an increasing awareness of the need to act. Just last week, President Obama acknowledged China in his speech to the joint session of Congress for having launched "the largest effort in history to make their economy energy efficient." Also last week, Greenpeace welcomed India's national climate plan's first step – a market mechanism that could phase-out 400 million incandescent bulbs until 2012. In December, last year Mexico set an aspirational target to cut in half its 2002 carbon emissions by 2050. And the Brazilian government released its National Plan on Climate Change. These are encouraging signs of action.

But the world has to do more and the world has to act together. Despite the action and efforts shown around the world, emissions continue to rise. We have to reverse this trend. Certainly, developed countries will have to show clear commitment and live up to their promises, and developing countries will need support when accelerating their action. To ensure that the world achieves the needed reductions, we need a strong agreement in Copenhagen. And we need to monitor and verify the efforts all across the world. We need to be sure that promises lead to action, that plans get implemented, that results live up to expectations.

The United States must show that it is will lead this effort. Only by doing so, we will collectively be able to win the fight against dangerous climate change.

Chapter 5

TESTIMONY OF CARTER ROBERTS, PRESIDENT AND CEO, WORLD WILDLIFE FUND, BEFORE THE SELECT COMMITTEE ON ENERGY INDEPENDENCE AND GLOBAL WARMING, HEARING ON "PREPARING FOR COPENHAGEN: HOW DEVELOPING COUNTRIES ARE FIGHTING CLIMATE CHANGE"

INTRODUCTION

Chairman Markey, Ranking Member Sensenbrenner, Members of the Committee: On behalf of World Wildlife Fund (WWF), I am pleased to present testimony to this committee. First, let me commend the Chairman and the Select Committee for its important work in bringing much-needed attention within the Congress to so many aspects of climate change. The many hearings held by this committee puts the Congress and the United States as a whole in a much better position to support the domestic legislation and international agreements necessary to respond to this global crisis. So thank you for your important leadership.

As the Congress works with renewed vigor on the critical question of how to construct a domestic framework to reduce emissions within the United

States, it is vital that we also remain focused on the need to work together with other nations, particularly developing countries and emerging economies, to produce a framework that ensures that global emissions hit their peak and begin to decline within the decade in order to limit overall temperature increase to below 2 degrees centigrade above pre-industrial levels. The impacts of climate change ignore our political borders; only a global solution will protect the people of the United States and all the nations of the world from the worst effects of climate change.

Conventional wisdom in Washington says that developing countries do not take climate change seriously, that emerging economies are not taking steps to reduce their emissions, and that these countries are an obstacle to reaching a new global agreement to stop climate change. Today, nothing could be further from the truth. Although it has become rare in these difficult times, I am here with good news: Developing countries 'get' climate change and they are taking action to reduce their emissions while constructively leading in the international negotiations.

But let's be clear, just as we are, developing countries are grappling with how best to meet the near-term energy needs of their growing populations, while also responding to the threats of climate change. And this is a much greater challenge for them than it is for us. Even the largest of these countries are poor and struggling by any measure we would use in the United States. For example, approximately 85% of the population of India lives on less than $2/day. This represents three times more people than the entire population of the US. And while their overall economies have grown in recent years, their gross domestic products are partially a function of very large populations, masking deep poverty. In truth, nearly half of the world's abject poor (living on less than $1/day) live in China and India alone; none live here in the United States.

Faced with these challenges, developing countries continue to struggle with how best to reduce emissions while responding to crushing poverty. They have not always succeeded in their attempts to reduce emissions and they are not in a position to make all of the necessary reductions on their own. This is no surprise. Emissions in the developing world continue to grow at a faster rate than in the industrialized world. This is also not a surprise. Many in the developing world are only now gaining access electricity and they understandably aspire to more of the basic conveniences that we take for granted. Moreover, as the world economy has become increasingly globalized, much of our demand for emission-intensive products, like beef, aluminum, lumber and cement is shifting to the developing world and along with the associated emissions.

What is a surprise, however, at least to some, is that these countries are doing a better job than the United States in taking ambitious action to reduce emissions, while leading the conversation on a new international climate agreement. These nations realize that future economic prosperity lies at the end of the road to a low-carbon economy. They hope to gain a competitive advantage in this new economy by acting now. And they have seen the early impacts of climate change on their people and their economies and realize there is no time to lose.

WORLD WILDLIFE FUND – A GLOBAL AND HISTORICAL VIEW

With operations in 100 countries and experience that stretches for nearly half a century, WWF has the geographic scope and historical perspective necessary to speak to past and current developing country actions and attitudes toward climate change. Since the late 1980's we have been working with local communities, governments, scientists, and businesses around the world to advocate for climate change solutions that will make the world cleaner, healthier, and safer. WWF's positions and perspectives on climate change are informed by deep technical expertise, and a global view based on knowledge of the domestic political situation in each relevant country and on-the-ground implementation projects at the local scale.

WWF's perspective is truly global. Our offices in China, India, Brazil, South Africa, Indonesia and elsewhere in the developing world are independent organizations, managed and led by local nationals with deep connections and understanding of their domestic context. On the international stage, WWF has been a mainstay in the climate negotiations as an official observer organization within the UN Framework Convention on Climate Change represented by a multinational delegation representing every major country in the negotiations. It is from this vantage point and with these combined voices that we provide testimony today.

International Climate Negotiations: A Changing Landscape

As with most conventional wisdom, the idea that developing countries are reluctant to take action to reduce greenhouse gas (GHG) emissions has some historical basis. By 1994, nearly every country in the world – including the United States – signed and ratified the UN Framework Convention on Climate Change (UNFCCC). Under that agreement, industrialized nations agreed to make commitments to reduce GHG emissions before any developing countries based on the recognition that industrialized country emissions were responsible for the lion's share of climate change.

Throughout much of the 1990s and into this decade, developing countries held fast to this bargain. Many, including China, argued strongly that they would take no action to reduce their emissions until after the United States – the world's largest historical emitter and the world's largest economy – took action. And so began over a decade of finger-pointing and strong rhetoric. For example, in 1997, a member of the Chinese delegation made clear that, at that time, China opposed making any emissions reductions. He said: "The position of the G-77 and China is clear -- no new commitments in whatever guise or disguise... [Developed countries] have to pay to the Earth the debt they owed since the Industrial Revolution."

And while this debate raged, emissions grew. Since 1995, U.S. emissions have increased approximately 14%. And although its emissions per capita remain quite low (99th in the world), in absolute terms, China now is the world's largest emitter of carbon dioxide (CO_2). Most importantly, our global emissions have continued to increase.

Overall emissions within these major emerging economies continue to grow at a faster rate than in the developed world and much of that is due to the globalization of trade. Today a larger and larger percentage of the emissions associated with the products we buy and the food we eat are generated outside of the United States. We continue to be a great driver for these emissions, but they are occurring overseas in developing countries.

Why Have Developing Countries Decided to Act?

Just as in the United States, the last few years have brought much greater awareness within the developing world of the great risks of climate change and, most importantly, how quickly the impacts would be upon us. As the evidence mounts that climate change resulting from human activity is well underway and that it is accelerating, developing countries have realized (more quickly than the United States, I'm disappointed to say) that the time for posturing is over, and the time for action has arrived. As importantly, developing countries have begun to understand that future economic prosperity will depend on investments in a clean, modern energy economy.

Seeing the Impacts

The ten warmest years on record have been 1997 through 2008 and during that time, the Intergovernmental Panel on Climate Change (IPCC) issued its third and fourth assessments (in 2001 and 2007). The fourth IPCC assessment report said that "warming is unequivocal" and that most of the observed increase in global average temperatures since the mid-20th century is very likely due to the buildup of greenhouse gases in the atmosphere resulting from human activity. The assessment also concluded that "[o]bservational evidence from all continents and most oceans shows that many natural systems are being affected by regional climate changes, particularly temperature increases."

The IPCC reports made clear that evidence of impacts from climate change on people and their activities is mounting for every region and more disruptive impacts are likely during the course of this century, including in the developing world. For example, the IPCC found:

- In Latin America, food security is likely to be jeopardized by declining productivity of key crops and livestock.
- Agricultural production could be "severely compromised" in many parts of Africa
- More than a billion people in Asia could be adversely affected by decreased freshwater availability

In the two years since the Fourth IPCC Assessment report was released, worrisome evidence has accumulated that climate changes will be larger and faster than the IPCC suggested in 2007.

A second factor motivating developing countries to respond to the threat of climate change is the fact that they already are experiencing climate change and its impacts. According to the IPCC, widespread changes have been observed in average temperatures, precipitation amounts, wind patterns and in extreme events such as droughts, heavy rains, heat waves and the intense tropical storms. These conditions – and their consequences – tend to intensify concerns about climate change and to stimulate efforts to respond to it. As President Obama's science adviser, John Holdren, said in his confirmation hearing several weeks ago, "the major developing country emitters like China and India have recognized that climate change is already harming them and it can't be fixed without them."

In China, for example, the worst drought in a half century is being experienced in eight provinces since November 2008, prompting China to declare its highest level of emergency in early February. Drinking water for over 4 million people has been affected, along with more than 24 million acres of cropland. In June and July 2007, it was the opposite extreme with devastating floods and landslides affecting seven provinces. When warm temperatures came early to China this year – with Nanjing experiencing the highest temperature in a century for the date – the chief forecaster for the Chinese National Meteorological Center (NMC) said "Spring has come early in some areas of East and Central China this year, and it's because of global warming,"

As we begin to see the impacts of climate change in the U.S., including extended drought patterns and wildfire seasons, some have asked whether it might be better to just accept climate change and pay to respond to the coming damage. For all nations, the economic impacts from climate change will likely soon outstrip any ability to simply pay for the impacts after-the-fact. But for poorer developing countries thinking about climate change in this way is not an option. These countries suffer from greater vulnerability to the effects of climate change due to their heavy dependence on natural systems and agriculture for subsistence. Moreover, they have limited capacity to respond and adapt to climate change given their limited financial resources. For nations with populations living at a subsistence level, even modest amounts of climate change are enough to risk crop failures, food shortages and loss of key water supplies. And for many developing countries, particularly small, island developing states, climate change poses a threat to their cultural survival. Wait-and-see is not an option.

Seeing the Opportunities

The impacts and the evidence of climate change do not tell the whole story of the turnaround by developing countries. Some of the answer comes from a basic recognition that reducing energy and reducing emissions is good for their economic prosperity. As we are beginning to understand in the United States, making short term investments in energy efficiency and modern technology results in reduced energy costs over the long term, more local jobs, and long-term economic growth. Countries just beginning to industrialize are looking to leapfrog our older, polluting approach in favor of newer, cleaner energy.

Moreover, the volatility of the price of foreign energy supplies such as oil and gas, have taught all countries the hard lessons of energy security. Dramatic swings in energy prices are especially problematic for poorer nations with fewer financial reserves. As a result home-grown energy supplies, starting with energy efficiency and including renewable power, offer a much firmer long-term foundation on which to build an emerging economy. At a time of greater economic uncertainty, wise investments in a sound economic future are more important than ever. (Of course, this is as true for the United States as any country. With Europe and the developing world beginning to lead on the new energy economy, the U.S. will continue to find itself at a competitive disadvantage.)

These concepts have become more greatly understood within the developing world; governments have instituted policies and the market place has responded. In some cases, emerging economies have learned these lessons better than we have. For example, a report issued last week by HSBC Global Research evaluated the economic stimulus plans implemented by various governments. Although in the United States the American Recovery and Reinvestment Act was rightly praised as including important investments in energy efficiency and renewable power, the U.S. stimulus act devoted only 12% of its funding for "investments consistent with a low carbon economy." Using the same criteria, China's stimulus plan was over three times more oriented towards promoting a low carbon economy (38%), while investing more money in these sectors in absolute terms.[1]

DEVELOPING COUNTRIES ARE TAKING ACTION

Whatever the motivations, the results are clear: In both the international negotiations and through action taken at home to reduce emissions, developing country governments have stepped down from the absolute demand that countries like the United States must act first to respond to climate change. They understand it is in their economic and national interest to stop waiting and move ahead. They are putting concrete proposals for mitigation on the table in the international negotiations, taking a constructive approach to climate and energy issues in bilateral and multilateral venues, and taking unilateral action to reduce greenhouse gas emissions at home.

Political Leadership

Despite bearing relatively little responsibility for the current impacts of climate change, emerging economies have determined that it is in their self interest to be part of the solution. In advance of the UNFCCC negotiations last December in Poznan, several key emerging economies offered comprehensive proposals to reduce their emissions, which included specific targets and timetables. Together with other recently-announced plans, these proposals marked a sea change in the international debate, breaking the log-jam of the previous decade where developing countries had refused to propose action until the United States made commitments to reduce emissions.

These proposals in many cases went beyond what we have been able to achieve in the United States and clearly indicate the leadership and firm commitment of developing counties to shift to low carbon economies. For example:

- South Africa established a plan that would halt its projected increase in emissions and produce a "peak and decline", a critical step towards changing the trajectory of future emissions towards stabilization – a step we hope the United States will take during this Congress. For South Africa, a country highly dependent on energy from coal, their peak and decline date of 2015-2020 was particularly ambitious.
- Mexico established an economy-wide plan to cut its projected emission *in half* by 2050 to be implemented through a cap-and-trade program.

- Brazil committed to reduce annual deforestation by 70% by 2018. Deforestation is the largest source of emissions in Brazil, and when deforestation is included, Brazil is one of the world's top emitters – making this target a significant step towards meeting global emissions trajectories that reduce the greatest impacts of climate change.
- India has committed to an economy-wide 20% increase in energy efficiency by 2016, while continuing its renewable energy program, one of the largest in the world.
- China has committed to reduce energy intensity of its economy by 20% by 2010, as well as an aggressive target to produce 10% of its primary energy through renewable sources by 2010 and 15% by 2020.

MITIGATION EFFORTS UNDERWAY

The seriousness with which these key nations have undertaken planning and targeting to reduce emissions is a significant step forward by itself. It has demonstrated recognition of the threat of climate change and an interest in transforming their economies towards a low carbon path, even where that would require new and significant changes.

But actions by emerging economies have gone beyond aggressive planning to actual emissions reduction against business-as-usual pathways. By studying indicators of progress in the energy sector, it is clear that developing countries have made notable advances. In two areas, the emissions intensity of economies and the use of renewable power, developing countries progress is equivalent to or exceeding progress in the United States.

Reducing Emissions Intensity

The carbon emissions intensity of an economy is expressed by the level of emissions per unit of economic output. This is a composite indicator determined by the combination of energy intensity and the fuel mix in a particular country. Emissions intensity levels are not linked to the size of a country's economy or population; a large or wealthy country may have a low GHG intensity or vice-versa. So this metric has greater policy relevance than absolute emissions. In other words, emissions intensity allows us to compare a country like India with nearly 1.2 billion people with a country like the United States with nearly 75% fewer people.

Let's be clear: simply reducing the emissions intensity of our economies is not enough. We must reduce the absolute levels of global emissions by at least 80% below 1990 levels by 2050 in order to reduce the greatest risk of dangerous climate change. Comparing the current carbon emissions intensity of various economies, however, demonstrates a nation's trend towards decarbonizing their economy (switching energy to lower carbon fuels, improving energy efficiencies, and/or restructuring economic activities). And so, this can be a useful way to compare how various nations are progressing towards a goal of absolute emissions reductions.

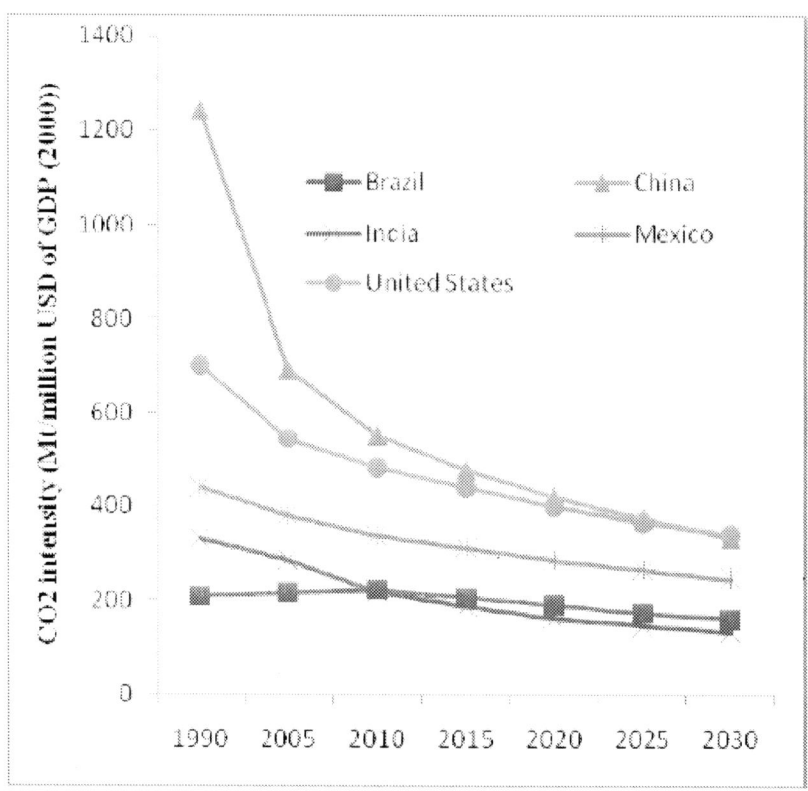

Source: Adapted from EIA Report#: DOE/EIA 0484-2008, Table 13.

Figure 1. Carbon Emissions Intensity of Selected Economies

Data from the International Energy Agency (IEA) show that China, India and Mexico have made good progress in de-carbonizing their economies (Figure 1), thus reducing emissions quite significantly for each unit of economic activity. In the United States, the de-carbonizing of the economy has been attributed greatly to the movement of higher-emitting sectors to the developing world. In the developing world, reducing the carbon intensity of the economy demonstrates, not the changing of industries, but the movement towards more modern, more efficient use of energy.

This movement is important and shows that policies being implemented by emerging economies are already working, resulting in real emissions reductions. For example, India has reduced the carbon intensity of its economy by over 35% since 1990. This reduction is related to India maintaining sustainable consumption patterns and enacting proactive policies to promote energy efficiency.

The reduction in the carbon intensity of the Chinese economy during this period has been even more dramatic (see figure 1). Since 1990, China has achieved remarkable energy efficiency. In 1990, the carbon intensity of the Chinese economy was about 1200Mt per unit of GDP; by 2005 that had been cut in half. Under its current 5 year plan, China has included a requirement to reduce the emissions intensity of its economy by an additional 20% below 2005 levels by 2010. If reached, this goal is estimated to reduce Chinese emissions by an additional 10% below business as usual levels.[2] China is making substantial progress toward this goal, reducing emissions intensity in the past three years by: 1.6% in 2006, 3.7% in 2007 and 4.3% in 2008. Although on pace to fall a bit short of this target, the ambition level and progress toward success are important steps indeed.

These improvements demonstrate the greatest reduction in emissions intensity of any major economy during the period and put China on track to match the emissions intensity of the United States in the near term. Although the decline in emissions intensity for China is not as fast as needed to offset China's rapid growth in energy consumption, the trend indicates a high prospect for China's transition to low carbon economy in the middle of the century. It also demonstrates the seriousness and effectiveness of Chinese policies to reduce emissions.

Renewable Energy Standards

As President Obama stated in his address to the nation on February 25th, "We know the country that harnesses the power of clean, renewable energy will lead the 21st century." Many key emerging economies apparently got the message long ago. One strong indicator of a commitment to low carbon energy is adopting a renewable energy standard (RES) and associated policies.

There are several forms of RES, including requiring an increase in total power generation capacity from renewable sources, an increase in the share of renewable energy in the primary energy supply, an increase in the share of electricity generated from renewable sources and an increase in the share of total energy consumption produced from renewable sources. Different developing countries have chosen various RES mechanisms (with some, like China, implementing many of these approaches simultaneously). In whatever form, an RES indicates a country's seriousness in replacing high-carbon fuels with ones that produce zero emissions. A successful RES reflects real emissions reductions below a business-as-usual case.

During the past several years, as the United States has debated whether to adopt any form of a renewable energy standard, Brazil, India, China, Mexico and other developing countries have begun implementing them with success (see Table 1):

- *China*: China has established an RES requiring 10% of its primary energy to be produced from renewable sources by 2010 and 15% by 2020. By 2006, 8% of its primary energy came from renewable sources and China is expected to meet these targets.
- *Mexico*: Mexico proposed a RES of 8% of electricity from renewable sources (excluding large hydro) by 2012. The Ministry of Energy has announced that the country is on track to meet that standard, driven mainly by installing wind power projects in the State of Oaxaca, which has an estimated wind power potential of over 10,000 MW.
- *Brazil*: Reflecting the highest percentage of renewable energy in the world, 46% of Brazil's primary energy comes from renewable sources, while over 75% of its electricity is produced from renewables. Brazil's high renewable share is largely driven by large hydro-electric facilities. Recognizing the need to shift to solar, wind, geothermal and small hydro, Brazil has implemented a RES of 15% from these sources by 2020.

- *India*: India has the 4th largest amount of installed wind power generating capacity in the world. In 2009, renewable energy power accounted for 8% of total power generation capacity in India; the country should meet and exceed its 10% RES by 2012. This success is a result of strong incentives from the government for enhancing renewable energy production capacity and power generation and the development of a framework for trading renewable energy certificates.
- *Philippines*: Another key developing country, the Philippines has the largest renewable target in the world, with a goal of producing 50% of its electricity from renewable sources by 2020. Philippines is currently the world's second largest producer of geothermal power and overall currently produces 33% of its energy from renewable sources.

Table 1. Illustrative renewable energy targets implemented in developing countries[3]

Country	Renewable target	Progress
India	10% by 2012[1]	India is on track to meet or exceed its renewable energy target, having already achieved 8% in 2009[1]
Philippines	50% by 2020[1]	Philippines is on track to achieve its target, and currently has 33% renewable energy in its power generation mix.
Brazil	15% by 2020[2]	Brazil's share of primary energy from renewables is currently 46%, among the highest in the world, relying heavily on large-scale hydro-electrical generation. This RES is focused on expanding wind, small hydro, and solar production from current levels of less than 4%.
Mexico	8% by 2012[3]	Mexico's Ministry of Energy expects to reach the country's goals, driven largely by new wind power projects in the State of Oaxaca
China	10% by 2010 and 15% by 2020[4]	By 2006 China had achieved 8% of its primary energy production from renewable energy, and is now scaling up wind and solar to meet these goals.

Table 1. (Continued)

Country	Renewable target	Progress
US	No National Renewable Target	A nationwide target is under discussion in both the House and the Senate. Current US percentage of electricity from renewables (not including large hydro-electric) is approximately 5% (2006).

1-Percent of total power generation in the country from renewable energy
2- Percent increase in the share of renewable energy in the primary energy supply
3- Percent of renewable electricity generation excluding large hydro
4- Percent of renewable energy in the primary energy supply

BRAZIL: AN EXAMPLE OF LEADERSHIP

As discussed, many developing countries are showing leadership in both the climate negotiations and by beginning to reduce their own emissions at home. Because one of the other panelists will focus specifically on China, this testimony will highlight another of these countries: Brazil. Brazil is the world's fifth-most populous country and the world's tenth-largest economy in GDP terms. When viewed at a human scale, however, the Brazilian economy is not as strong: in GDP per capita (PPP), Brazil ranks 82nd in the world.

Although no country has a perfect record in responding to climate change, Brazil has become a leader in reducing the emissions intensity of its economy, in generating renewable power and, perhaps most importantly, in seriously addressing emissions related to deforestation. As previous sections of this testimony discussed the first two of these, this section will discuss the third.

Although often forgotten as a major source of greenhouse gas emissions, deforestation is actually the second largest source of emissions by sector, producing approximately 20% of global emissions – more than every car, truck, plane, train and boat on the planet. In the developing world, deforestation-related emissions constitute an even larger share of the total. For example, when deforestation-related emissions are included, Brazil ranks 7th in the world in absolute emissions, despite producing nearly 50% of its electricity from sources that do not emit GHGs. These high emissions are largely associated with deforestation, which accounts for about 75% of the country's emissions.

Reducing emissions from deforestation in a lasting way requires substantial upfront investment in building monitoring capacity, improving measuring and accounting systems, engaging in extensive land tenure reforms to ensure that local landowners are properly compensated and increasing investment in law enforcement. These kinds of investments in a national program demonstrate a commitment to ensure that forest programs result in reduced GHG emissions. Absent this type of investment, project-level deforestation reduction activities may not provide reliable benefits to the climate.

The required investment is substantial, but the government of Brazil has committed to building this capacity to reduce deforestation-related emissions, including:

- Establishing 148 protected areas covering 620,000 km2 from 2003-2007. Many of these new protected areas are located in zones under high deforestation pressure.
- Developing and implementing one of the most sophisticated forest change tracking systems in the world, based on remote sensing methods and linked in to land management databases in state-level governments. This system is so widely regarded that it is being made available to other governments.
- Stepped-up enforcement against illegal logging, deforestation and other environmental crimes.
- Prohibiting financing for landholders without clear tenure or in breach of environmental laws.
- Accelerated land reform to establish clear tenure rights in areas subject to intensive social conflict.
- Developing a legal framework for forestry concessions in public forests.

These efforts have helped substantially reduce deforestation in the Brazilian Amazon by 56% in since 2004. This alone represents a decrease of *1.3 billion tons of CO2 emissions* in relation to the previous 4-year period, or nearly 20% of the U.S.'s current annual emissions of CO2-e (7.0 billion tons in 2006). Building on these actual reductions, in December the Brazilian government announced a new target of reducing deforestation by 70% below 2006 levels by 2017. This would avoid *4.8 billion tons of CO2 emissions* – equivalent to over two-thirds of current annual emissions in the United States.

Meeting this new, ambitious goal will not be easy and Brazil cannot do it alone. But its commitment to making the necessary early investments and continuing to press for even greater reductions shows Brazil to be a leader. And it further helps to replace the old conventional wisdom about developing countries with a new reality: these nations are taking action and looking to partner with the rest of the world to do even more.

CONCLUSION

For over a decade, the United States has failed to take serious action to address climate change because of the perception that major emerging economies were not acting to reduce their own emissions. This justification for inaction was always flawed, as it ignored the seriousness of the problem, our historical responsibility, our commitments made under the UNFCCC and the power of American leadership. But flawed or not, today it is gone. Developing countries, in particular the major emerging economies, are taking action to reduce their emissions, even in the absence of U.S. leadership and action. We must quickly follow suit.

These actions are but the first important steps on a long journey. Due to increasing populations and the beginning of industrialization, emissions trends indicate that the majority of new emissions will be produced in the developing world in the coming decades. As they grapple with both climate change and desperate poverty, developing countries will need our help to fully make the transition towards low-carbon economies. Based on our historical responsibility for the climate crisis and our greater economic capacity, this help is justified.

As domestic cap-and-trade legislation is designed and debated within the Congress, it is important to keep this international context in mind. Climate change is a global problem and its impacts can only be slowed through a global response. In addition to helping middle- and low-income Americans transition to a clean energy future, some revenues from a cap-and-trade system are needed to provide predictable finance for emissions reductions and adaptation needs within the developing world. Only by helping developing countries continue to move toward low carbon pathways and reducing emissions from tropical deforestation, can we bring global emissions under control. Importantly, this is not a zero sum game. Improving the global market for clean energy technologies will help spur greater advancements throughout the industry, which will also help the transition to low-carbon pathway in the United States.

Their recent actions show that these emerging economies will be good partners in this global effort. They have demonstrated their serious commitment to addressing climate change and have sent a strong message to the world that they are ready for a new era of international cooperation. They have taken actions and developed plans to begin to decouple their economic growth from their greenhouse gas emissions, so that they can grow sustainably without disrupting the climate system.

The United States should follow their example and begin to assume responsible leadership both at home and abroad to address the climate crisis.

End Notes

[1] A Climate for Recovery: The Color of Stimulus Goes Green, HSBC Global Research (25 Feb. 2009) at 2.
[2] Climate Change Mitigation Measures in the People's Republic of China, Pew Center on Global Climate Change (April 2007)
[3] Data primarily adapted from Renewables 2007: Global Status Report, REN21:Renewable Energy Policy Network for the 21st Century (2007).

Chapter 6

TESTIMONY OF BARBARA A. FINAMORE, SENIOR ATTORNEY AND CHINA PROGRAM DIRECTOR, NATURAL RESOURCES DEFENSE COUNCIL, PRESIDENT, CHINA-U.S. ENERGY EFFICIENCY ALLIANCE, BEFORE THE SELECT COMMITTEE ON ENERGY INDEPENDENCE AND GLOBAL WARMING, HEARING ON "PREPARING FOR COPENHAGEN: HOW DEVELOPING COUNTRIES ARE FIGHTING CLIMATE CHANGE"

Chairman Markey, Ranking Member Sensenbrenner, and distinguished Members of the Committee, it is my pleasure to be here with you today to discuss China's national greenhouse gas mitigation efforts and achievements. I applaud the committee for calling a hearing on the vitally important topic of how developing countries, including China, are already taking action to fight climate change.

Beginning with its Eleventh Five-Year Plan, which covers 2006 to 2010, China has recognized that it must reduce its rapid growth in energy demand and greenhouse gas emissions and accordingly has embarked on what

President Obama in his speech to Congress last week called "the largest effort in history to make their economy energy efficient." It is important that the United States understand what measures China is taking to reduce its greenhouse gas emissions as well as how the United States can strengthen its engagement with China on climate change, because China and the United States together are the two countries that can have the greatest impact on mitigating climate change.[1] The Chinese viewed Secretary of State Clinton's recent visit to Beijing and her message of cooperation extremely favorably and are eager to find areas for mutual cooperation.

I am a Senior Attorney and Director of the China Program for the Natural Resources Defense Council (NRDC), and have worked on China energy and environmental issues for nearly twenty years. NRDC is a nonprofit environmental organization with a staff of nearly 400 lawyers, scientists and policy experts, including a staff of 25 working full time in Beijing on energy, climate, and environmental governance issues. Over the last twelve years, recognizing the importance of China to the global environment, we have been working with the Chinese government to help reduce China's CO_2 emissions by developing national energy codes and standards for buildings and equipment, promoting demand side management (DSM) energy efficiency programs and advanced energy technologies, and focusing on ways to improve environmental enforcement and governance. I am also the President of the China-U.S. Energy Efficiency Alliance, a nonprofit organization that promotes technical exchanges between U.S. and Chinese government officials, utilities and energy experts to help China design and implement large-scale DSM energy efficiency programs targeted at China s industrial sector.

THE ORIGINS OF CHINA'S PRESENT GREENHOUSE GAS MITIGATION EFFORTS

China is currently pursuing an aggressive and ambitious greenhouse gas mitigation program—a result of its recognition that its present development model is unsustainable and that climate change is likely to have serious impacts on its agricultural productivity and water resources (causing droughts in the north and flooding in the south and on its coasts), increase the incidence of extreme weather events, and lead to deterioration of its forests and other natural ecosystems. China is also keenly aware of the intimate connection between its enormous growth in energy demand and its energy security, and the serious public health and environmental damage caused by emissions of pollutants such as SO_2, NOx, particulate matter and mercury from its coal-dominated energy system.

China's greenhouse gas mitigation efforts are reflected in its National Assessment Report on Climate Change (December 2006), National Climate Change Action Plan (June 2007), and a Climate Change White Paper (October 2008), but have their roots in China s response to the tremendous and unexpected surge in energy growth that occurred beginning in 2002. From 1980 to 2000, China's GDP quadrupled, but energy demand only doubled from 603 to 1,386 mtce ("million tons of coal equivalent") as a result of Chinese policymakers' emphasis on energy efficiency. In other words, energy grew at half the rate of GDP growth. Between 2000 and 2005, however, China's primary energy consumption skyrocketed from 1,386 to 2,225 mtce, an annual rate 1.5 times faster than the growth in GDP.[2] This sudden increase in energy demand and hence greenhouse gas emissions was not predicted by either international or domestic energy experts, and led China to rapidly increase its thermal power plant capacity to meet its energy needs. In 2006, for example, China added 90 GW ("gigawatt")[3] of coal-fired power capacity this addition alone is enough to emit over 500 million tons of CO_2 per year for 40 years.[4] To put this in comparison, the entire European Union s Kyoto reduction commitment is 300 million tons of CO_2.[5] The rapid growth in energy demand occurred primarily because of an increasing dominance of heavy industry in China s economic structure—i.e., cement, iron and steel, and chemicals—which overshadowed improvements in energy efficiency.[6] It is this rapid growth in energy demand that resulted in China overtaking the United States as the largest greenhouse gas emitter some time in 2006 or 2007.[7]

Recognizing the need to rein in energy growth, China's leaders set out in the Eleventh Five Year Plan a goal of reducing energy intensity (energy consumption per unit of GDP) 20 percent from 2005 levels by 2010, *i.e.*, a 4 percent reduction per year. They also set a target of increasing the share of renewables in the energy mix to 10 percent by 2010 and 15 percent by 2020. If China succeeds in reducing its energy intensity by 20 percent by 2010, it will avoid emitting approximately 1.5 billion tons of CO2,[8] constituting the largest single greenhouse gas mitigation program by any country. I will first address China's efforts to reduce energy demand, then discuss their efforts to reduce the carbon intensity of their energy supply.

CHINA'S EFFORTS TO IMPROVE ENERGY EFFICIENCY AND REDUCE ENERGY DEMAND

China appears to be making some progress in reaching its energy intensity goals. After reducing energy intensity by only 1.23 percent in 2006, it reduced energy intensity by 3.66 percent in 2007 and 4.59 percent in 2008. It accomplished this primarily through economic restructuring and a renewed emphasis on energy efficiency, although the global economic downturn also played a role starting in the last quarter of 2008. Major initiatives include:

- Replacing smaller, less efficient power plants and closing backwards production capacity,[9] slowing the expansion of high energy-consuming industries through the elimination or reduction of export tax rebates for energy intensive products, and using differential pricing of electricity and a Green Credit policy to encourage more efficient enterprises and limit or shut down less efficient enterprises.[10] China is also encouraging growth in the service and high-tech industries.
- A renewed emphasis on energy efficiency, particularly in the industrial sector, which accounted for 77 percent of delivered energy use in China in 2005.[11] One particular project of note is the "Top 1000" program, started in April 2006 to improve the energy efficiency of the top 1,000 energy consuming enterprises in nine sectors,[12] with a goal of saving 100 mtce by 2010. These 1,000 enterprises alone constituted 33 percent of national energy consumption and 47 percent of industrial energy consumption in 2004, and represented approximately 43 percent of China's CO_2 emissions in 2006.[13] Although data is limited, a preliminary assessment concluded that the Top 1000 program is on track to meet or surpass its target of saving 100 mtce per year, which would translate into a reduction in CO_2 emissions of 300 to 450 million tons, and could constitute 10 to 25 percent of the savings necessary to meet China s 20 percent energy intensity reduction target.[14]
- Funding energy efficiency at the national and provincial level. The central government allocated 23.5 billion RMB ($3.4 billion) in 2007 and 41.8 billion RMB ($6 billion) in 2008 to promote energy efficiency and reduce emissions.[15] A portion of this funding is used to reward enterprises that can demonstrate aggregate savings of 10,000

tons of coal equivalent per year from energy conservation projects by providing 200-250 RMB ($29-36) for every ton of coal saved.[16] Provincial-level energy efficiency funds also exist; for example, Shandong province has initiated a 2.13 billion RMB ($304 million) fund for local enterprises.[17]

- Beginning to implement provincial and municipal demand side management programs, based on experience in states such as California, in which utilities or another regulated party uses technical assistance, funding (such as a system benefits charge fund) and information programs to reduce peak load and overall energy demand through large-scale investments in energy efficiency. The NRDC and China-U.S. Energy Efficiency Alliance have established a pilot program that has avoided the need to build 300 MW of electric capacity in Jiangsu province, eliminating 1.84 $mtCO_2e$.[18] A World Bank study concluded that with the proper policies and incentives, DSM programs could reduce electricity needs by 220 terawatt hours[19] and avoid the need to build more than 100 GW of electric capacity by 2020.[20] Tapping even half of this potential would reduce coal consumption by about 37 mtce in 2020, avoiding 93 million tons of CO_2 emissions.

- Continuing to develop and implement building energy codes, appliance and equipment energy efficiency standards and labeling programs, and stricter vehicle fuel efficiency standards. China is building 2 billion square meters of floor space each year, half of the world's total. As more and more Chinese move into cities and begin to use modern conveniences and personal automobiles, maintaining efficiency standards will be crucial to slowing greenhouse gas emissions. Between 2000 and 2020, improved efficiency in electric appliances and gas water heaters is projected to reduce carbon emissions by more than 1.1 billion tons of CO_2.[21] However, monitoring and enforcement of these standards will be crucial to ensuring that the potential GHG reductions from these programs are indeed achieved.

- Launching a rebate program last April to subsidize the purchase of energy efficient light bulbs, offering a 30 percent subsidy on wholesale purchases and a 50 percent subsidy on retail sales. Some local governments offered additional subsidies of up to 40 percent. Lighting now accounts for about 12 percent of China's total electricity consumption, and using energy-saving bulbs could cut such power consumption by 60 to 80 percent. By the end of January 2009, 62 million energy-saving light bulbs had been sold under the subsidy program, which will help save 3.2 billion kWh of electricity annually and eliminate 3.2 million tons of CO_2 emissions. China announced last week that it will double the size of the program in 2009, subsidizing 100 million energy-efficient light bulbs this year. China is also beginning to work on a program to phase out the more than 1 billion ordinary bulbs that the country consumes every year.

- Raising fuel economy standards from 36 to 43 mpg this year and instituting graduated sales taxes favoring smaller cars, in order to slow growth in oil consumption fueled by the rapid expansion in personal vehicles.[22] China has also begun to bring its oil prices more in line with international markets, which would reduce demand over the long term. In early 2009, Beijing prohibited the driving of all heavily polluting yellow-label vehicles (which account for 10 percent of the total number of motor vehicles but 50 percent of emissions) within the city limits, and provided cash rebates to help ease the transition. The Chinese government also announced two weeks ago that it will offer cash rebates ranging from 50,000 RMB ($7,353) for small hybrid passenger cars to 600,000 RMB ($87,719) for large, fuel cell powered commercial buses in 13 major cities, including Beijing and Shanghai. China plans to put 60,000 new-energy vehicles for trial runs in 11 cities by 2012 for public transportation and public services. And the Ministry of Railways just signed a purchase agreement valued at 27 billion RMB ($3.95 billion) to purchase 500 clean and energy-efficient locomotives to replace diesel-powered engines on various lines. Replacing one diesel locomotive with an electric locomotive is equivalent to eliminating emissions from 4,000 vehicles.

- Finally, the government is raising public awareness of the need to "save energy and reduce emissions" (*jie neng jian pai*), has amended the energy conservation law, and is using improvements in energy efficiency as one measure by which government officials' performance is evaluated. A recent nationwide public opinion survey found that three out of every four Chinese citizens, or 76 percent, believe that environmental problems in China are "very serious" or "relatively serious."

Although China has made significant progress in improving its energy efficiency, much more can be done. China's energy intensity is currently four times that of the US and nine times that of Japan. According to McKinsey Global Institute estimates, if China were to pursue all cost-effective energy efficiency options, it could reduce energy use by approximately 1,050 mtce in 2020.[23] By doing so, it could cut its projected energy demand by about 23 percent and its CO_2 emissions by at least 20 percent from a business as usual scenario.[24] Many of these investments in energy efficiency are cost-effective, and the International Energy Agency estimates that on average every additional $1 spent on more efficient electrical equipment, appliances and buildings avoids more than $2 in investment in electricity supply.[25] It is estimated that China will require investments of 150-200 million RMB ($21-29 billion) per year to reduce the growth rate of energy demand to half the growth rate of the economy over the next 15-20 years.[26]

Further progress in reducing energy demand will depend on reforming incentives in the electric industry, so that grid companies and utilities are rewarded for improving energy efficiency (i.e., decoupling), and engaging with commercial banks and the small but growing Energy Service Company market to ensure that energy efficiency investments can be a sizable and sustainable market. There is also a need to increase technical capacity in energy auditing and energy efficiency retrofit design and implementation. Implementation of all of these efforts will require a sustained effort to monitor and enforce energy efficiency standards in industry, power plants, buildings, appliances, equipment and automobiles.

China's Efforts to Reduce the Carbon Intensity of its Energy Supply

China has also sought to reduce the carbon intensity of its energy supply by closing down smaller, inefficient thermal power plants and increasing the share of less carbon-intensive sources of energy, notably hydropower, wind and nuclear:

- At the end of 2008, China had a total installed electric power capacity of 792 GW, constituting 76 percent thermal power capacity, 22 percent hydropower, 1.6 percent wind, and 1.1 percent nuclear. In 2008, it expanded its total thermal power capacity by 66 GW, or about two new 600 MW power plants per week. This is consistent with the pace of expansion in recent years, which saw thermal power additions jump from an increase of 15 GW in 2004 to 86 GW in 2005, 93 GW in 2006 and 70 GW in 2007. China also expanded hydropower by 20 GW and windpower by 6.4 GW in 2008. In terms of actual electricity generated, China generated a total 3,443 TWh in 2008, comprised of 2,779 TWh of thermal (80.1%), 563 TWh of hydro (16.4%), 68.4 TWh of nuclear (2%) and 12.8 TWh of wind (0.4%).

- China's thermal power expansion, however, is occurring through replacement of smaller, less efficient power plants with larger, more efficient plants. China shut down 34 GW of small, inefficient plants from 2006-08, and plans to close another 31 GW of inefficient plants during the next three years. This has improved average efficiency from about 370 grams of coal per kWh in 2005 to 349 grams of coal per kWh in 2008. China is also pursuing cleaner thermal power generation technologies such as combined heat and power, integrated gasification combined cycle (IGCC) and carbon capture and storage (CCS). GreenGen, a joint venture established by Chinese utilities, is building China s first IGCC power plant in Tianjin, which is slated to come online with 250 MW capacity in 2010 and expand to 650 MW with CCS by 2020. The Chinese utility Huaneng Group started a pilot CCS project in Beijing last summer.
- China passed a Renewable Energy Law in 2005 and a Medium and Long-Term Development Plan for Renewable Energy in 2007 to encourage the growth of renewables. The renewable energy plan calls for the share of renewable energy in primary energy to reach 10 percent by 2010 and 15 percent by 2020. To help meet these targets, China has regulations that encourage the construction of renewable energy facilities and offer financial incentives and reduced taxes for renewable energy projects, including loan discounts and a feed-in program.
- Wind capacity doubled in 2008 from 5.9 GW to 12.3 GW, thus passing China's goal of 10 GW wind capacity by 2010. China plans to continue its rapid expansion of windpower and to improve the quality of its domestically manufactured wind turbines and connection with the grid. China s installed solar capacity is small but growing; it is planning to build a 10 MW solar PV power plant in Dunhuang, which would be the largest in the country. China is the world s largest manufacturer of solar PV panels, although almost all of this is exported. China also produces 80 percent of the world's solar water heaters, which make up 20 percent of its water heating units.

- China presently has 11 nuclear reactors with 9 GW of capacity, accounting for over 1 percent of its energy mix. It is likely to raise its target from 4 percent to 5 percent of energy production by 2020, or about 60 to 70 GW total capacity.

Finally, let me mention a few mitigation measures that do not fall neatly into the categories of energy efficiency or energy supply. China has afforestation efforts and forest management efforts aimed at raising forest coverage to 20 percent by 2010. It is exploring the idea of eco-cities and smart growth and increasing the use of mass transit. It has joined the Methane to Markets Partnership to better utilize coal bed methane. In addition, China s 4 trillion RMB ($585 billion) economic stimulus package includes 600 billion RMB ($88 billion) for building intercity rail lines, 476 billion ($70 billion) for new electricity grid infrastructure and 350 billion RMB ($50 billion) for energy efficiency and environmental protection projects.

In sum, China is working aggressively to improve its energy efficiency and to reduce the carbon intensity of its energy mix. The speed with which its economy is growing means that it faces a challenging task, but a sustained and sizeable effort to reduce its energy demand and the carbon intensity of its energy supply could result in a substantial reduction in the growth of its greenhouse gas emissions. According to a recently issued McKinsey study, if China pursued energy efficiency to the full extent possible and cut coal to 34 percent of its power supply, it could nearly cut in half its projected greenhouse gas emissions in 2030.[27]

China is taking enormous steps to reduce its impact on climate change and it is likely to continue and possibly intensify these efforts in the future. It has devoted significant economic and political resources to achieving the targets it has set for itself and demonstrated a willingness to pursue results. The United States can take China's mitigation actions to date as a strong signal that it intends to take concrete and meaningful steps to address climate change in the future.

I thank the committee for inviting me to participate in this hearing and look forward to answering any questions you may have.

End Notes

[1] For the Committee's benefit, I have attached to this written testimony a set of recommendations by the Natural Resources Defense Council on "Strengthening US-China Climate Change and Energy Engagement."

[2] World Bank, *Sustainable Energy in China* (2006), pp. 11.

[3] One gigawatt is equivalent to 1 billion watts.

[4] Statement of Stephen Chu, Director, Lawrence Berkeley National Laboratory, before the U.S. Senate Committee on Finance, March 27, 2007.

[5] *Id.*

[6] Trevor Houser, Testimony before the U.S.-China Economic and Security Review Commission, June 14, 2007.

[7] Netherlands Environmental Assessment Agency, China now no. 1 in CO2 emissions; USA in second position, June 19, 2007, available at www.pbl.nl.

[8] Jiang Lin, Nan Zhou, Mark Levine, and David Fridley, "Taking out 1 billion tons of CO2: The magic of China's 11th Five-Year Plan?", Energy Policy 36 (2008): 954-970.

[9] In 2007, for example, China closed 14.38 GW of small thermal power plants, 46.59 million tons of iron smelting capacity, 37.47 million tons of steelmaking capacity and 52 million tons of cement production capacity.

[10] Lynn Price, Xuejun Wang and Jiang Yun, *China's Top-1000 Energy-Consuming Enterprises Program: Reducing Energy Consumption of the 1000 Largest Industrial Enterprises in China*, LBNL-519E, pp. 9.

[11] EIA, *International Energy Outlook 2008*, available at http://www.eia.doe.gov/oiaf/ieo/world.html.

[12] The nine sectors are iron and steel, non-ferrous metal, chemicals, petroleum/petrochemicals, construction material, textiles, paper, coal mining and power generation.

[13] L. Price et al., *China's Top-1000 Energy-Consuming Enterprises Program*, pp. 18.

[14] *Id.*, pp. 27.

[15] *Id.*, pp. 8.

[16] *Id.*

[17] *Id.*

[18] Million tons of CO_2 equivalent.

[19] A terawatt is equivalent to 1 trillion watts.

[20] Zhaoguang Hu, David Moskowitz, and Jianping Zhao, *Demand Side Management in China's Restructured Power's Industry* (December 2005), World Bank Energy Sector Management Assistance Program.

[21] Mark D. Levine and Nathaniel T. Aden, *Global Carbon Emissions in the Coming Decades: The Case of China* (2008), Annual Review of Environment and Resources, pp. 32.

[22] 9.4 million cars were sold in China in 2008.

[23] Based on estimates by the McKinsey Global Institute in *Curbing Global Energy Demand Growth: The Energy Productivity Opportunity* (2007), pp. 34.

[24] McKinsey Global Institute, *Leapfrogging to Higher Energy Productivity in China* (2007).

[25] International Energy Agency, *World Energy Outlook 2006*.

[26] L. Price et al., *China's Top-1000 Energy-Consuming Enterprises Program*, pp. 8.

[27] McKinsey & Company, *China's Green Revolution: Prioritizing Technologies to Achieve Energy and Environmental Sustainability* (2009).

Chapter 7

TESTIMONY OF NED HELME, PRESIDENT, CENTER FOR CLEAN AIR POLICY, BEFORE THE SELECT COMMITTEE ON ENERGY INDEPENDENCE AND GLOBAL WARMING, HEARING ON "PREPARING FOR COPENHAGEN: HOW DEVELOPING COUNTRIES ARE FIGHTING CLIMATE CHANGE"

Mr. Chairman, Ranking Member Sensenbrenner and Members of the Committee: I would like to thank you for the opportunity to testify before you today. My name is Ned Helme and I am the President of the Center for Clean Air Policy (CCAP), a Washington, DC and Brussels-based environmental think tank with on the ground programs in New York, San Francisco, Mexico City, Beijing, Jakarta and many other places.

Since 1985, CCAP has been a recognized world leader in climate and air quality policy and is the only independent, non-profit think-tank working exclusively on those issues at the local, national and international levels. CCAP helps policymakers around the world to develop, promote and implement innovative, market-based solutions to major climate, air quality and energy problems that balance both environmental and economic interests.

CCAP is actively working on national legislation in the United States (U.S.) and is advising European governments as well as developing countries such as China, Brazil, and Mexico on climate and energy policy. Our behind the scenes dialogues educate policymakers and help them find economically and politically workable solutions. Our Future Actions Dialogue provides in-depth analyses and a "shadow process" for climate negotiators from 30 nations from around the world to help them develop the post-2012 international response to climate change. We also facilitate policy dialogues with leading businesses, environmental groups and governments in the European Union and U.S. on designing the details of future national and transatlantic climate change mitigation, adaptation and transportation policies.

My testimony responds to three questions posed by the Committee:

- How will the climate negotiations in Copenhagen in December 2009 differ from those at Kyoto in 1997?
- What are the key elements of the Bali Action Plan and how will they affect developing country expectations and the negotiations?
- How will leadership by the U.S. influence the emerging economies to scale up their efforts on mitigation?

We are at an exciting and very productive time in the negotiations to reach a global agreement on the framework for reducing global greenhouse gases (GHG) to a level that will prevent dangerous anthropogenic interference with the climate system. Never before have the stakes been so high or the opportunity so great to reach a globally acceptable deal that involves both the developed and the major developing economies.

In my time today, I would like to emphasize a few key points:

- The Bali Roadmap is the breakthrough developed countries have been waiting for that makes the negotiations in Copenhagen very different from those in 1997 and will bring meaningful developing country actions into the agreement.
- Developing countries are taking action already and are prepared to take additional measurable, reportable and verifiable actions.
- U.S. willingness to propose and enact a meaningful domestic national emissions reduction target is a linchpin for a successful outcome in Copenhagen.
- The objective in Copenhagen is to agree on new GHG reduction goals along with a new architecture to govern developing country action in the post-2012 framework, and

- The U.S. can successfully protect domestic, internationally competitive industries from job losses associated with a carbon program, while also creating incentives for developing countries to take greater action than they have underway already.

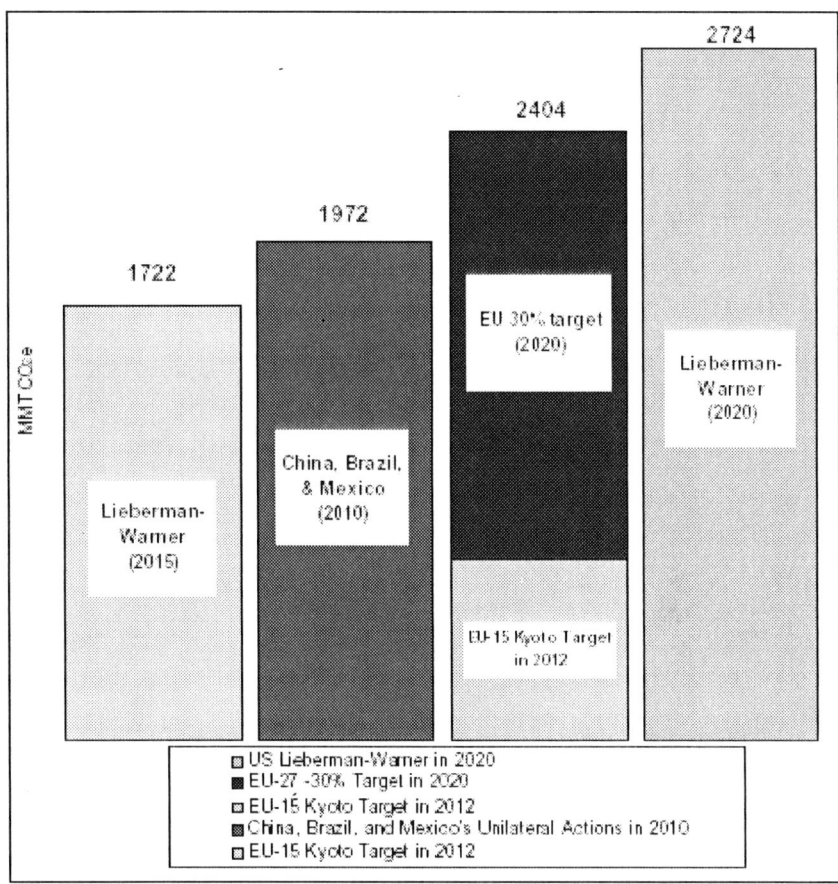

Figure 1. Emissions reductions from BAU for full implementation of proposed measures (CCAP, 2009)

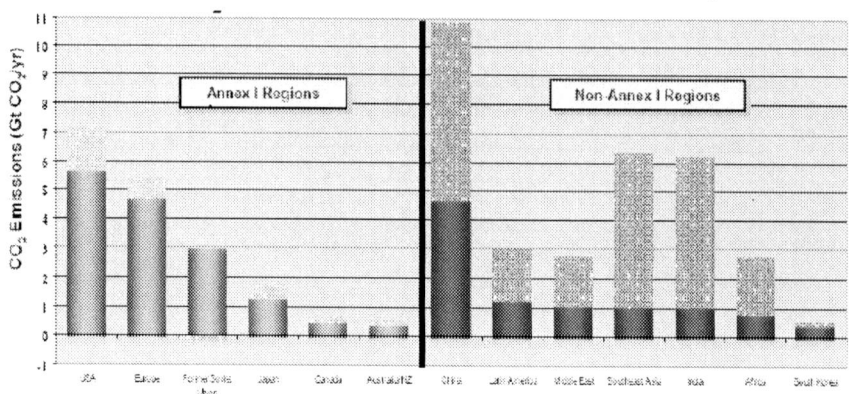

Figure 2. Fossil Fuel and Industrial Process CO_2 Emissions by Region in 2000 (solid bars) and 2050 (checkered bars) (U.S. Climate Change Science Program. 2007. *Scenarios of Greenhouse Gas Emissions and Atmospheric Concentrations*; MINICAM Results).

Our extensive policy work in key developing countries has shown that they are doing more to reduce the growth in their emissions than conventional wisdom here in the United States would suggest. China, Brazil and Mexico have already put in place national laws that collectively, if fully implemented, will reduce the projected growth in emissions by more aggregate tons in 2010 than the reductions the Lieberman-Warner bill (S. 2191 of the 110th Congress) was projected to achieve by 2015 and by almost as many tons as the European Union's 30 percent reduction pledge for 2020 (Figure 1).

Nevertheless, the outlook for developing country CO_2 emissions growth remains substantial in the aggregate and as a percentage of global emissions (Figure 2). In 2000, developed country emissions from fossil fuels and industrial processes were roughly 40 percent of global emissions. By 2050, developing country emissions are expected to grow to 64 percent of global emissions.

The negotiations going into Copenhagen are notably different than the 1997 Kyoto negotiations because we now have in place the Bali Action Plan, which the U.S. and other developed and developing countries agreed to in December 2007. The Action Plan builds on the key principle in Article 3 of the United Nations Framework Convention on Climate Change (UNFCCC), "The Parties should protect the climate system...on the basis of equity and in accordance with their *common but differentiated responsibilities and respective capabilities*." However, it goes much farther and establishes for the first time that the negotiation process will cover both developed and developing country actions to mitigate climate change. It also importantly sets up much stronger accountability by calling for developing countries to consider: "Nationally appropriate mitigation actions ... in the context of sustainable development, supported and enabled by technology, financing and capacity-building, in a measurable, reportable and verifiable manner". In effect, both the actions and the support are to be measured, reported, and verified.

In keeping with this new framework, the discussions since Bali have begun to define a menu of options for what are referred to as "Nationally Appropriate Mitigation Actions" (NAMAs). It is expected that each developing country will choose those actions that make the most sense for its own circumstances, just as we will do in the U.S.

NAMAs could take three distinct forms: *unilateral actions* that developing countries will take on their own without any assistance; *conditional actions* they will take conditioned on receiving financial and technology assistance from developed countries; and *emission credit generating policies* — where credits may be earned and sold in the international market if the country exceeds the goal it has set.

Although all developing countries will be encouraged to implement actions, the main focus appropriately will be on the six to ten largest emitting countries in the developing world which are responsible for 80-90 percent of the emissions in key industrial sectors. Reaching agreement on specific actions in these countries and on the support for those actions from developed nations will be the key to the Copenhagen agreement.

The Kyoto Protocol has long been criticized in the U.S. and elsewhere for the fact that it does not require explicit emission reductions by developing countries. Instead, it rewards developing countries who implement specific emission-reducing projects with emission credits through the Clean Development Mechanism (CDM) that can be sold to developed countries or to companies and individuals within such countries. These credits can be used to meet domestic carbon reduction requirements in developed nations. In effect, these reductions are paid for by developed nations.

The Kyoto Protocol does not contain any explicit system for recognizing actions taken by developing countries to reduce GHG emissions outside the CDM. One of the tests of any agreement in Copenhagen will be whether it creates a system for recognizing unilateral actions by developing nations to reduce their emissions that constitute their contribution toward protecting the climate. A large portion of the more than 2 billion tons of projected reductions in emissions growth by China, Brazil, and Mexico that I detailed for you earlier in Figure 1 of my testimony are unilateral reductions that contribute to protection of the climate, not reductions that generate credits for sale to developed nations. These unilateral actions are one form of a NAMA. Negotiators have proposed creating a formal registry in the UNFCCC that will record these and other NAMAs proposed by developing nations.

Recent actions by key developing countries give us a sense of what some of these actions or NAMAs might look like. For example, in Poznan, Poland, in December 2008, Mexico took a significant step, announcing its plans to set a national aspirational goal to reduce absolute emissions by 50 percent below 2000 levels by 2050. It also announced plans to set emission goals for four key industrial sectors — cement, steel, aluminum and electricity — and to achieve these goals through a domestic cap and trade program. It suggested an initial reduction target that it would undertake unilaterally in each sector and suggested that each sectoral target could be made more stringent if developed nations provided focused loan support (to overcome domestic financing barriers) in the post-2012 agreement. Mexico has also created and financed its own Energy Transition Fund of three billion Mexican pesos a year for three years (about $210 million annually) to provide incentives for more aggressive emissions reduction activities.

There are two key elements here that distinguish this from today's CDM approach: 1) the support for a more stringent sector-wide policy involves loans, not full payment for the incremental emissions reductions, and 2) it does not involve any generation of offset credits for developed nations in meeting the new more stringent target. All of these reductions will help reduce global aggregate emissions to safe levels rather than replacing or offsetting required reductions by developed nations. Offset credits would be generated only if the sector (e.g. Mexican oil refining) reduces its emissions in aggregate below the sectoral cap level. The heart of this program is then to generate a Mexican net contribution to the protection of the climate.

China also has taken bold action to reduce emissions. The government released its climate plan in 2007 and has set an aggressive goal to reduce its energy use per unit of GDP by 20 percent between 2006 and 2010. In the plan's first year in 2006, China fell short of its 4 percent per year goal, but in 2007 and 2008 it has reached the aggregate 8 percent reduction for those two years. If fully achieved, this goal alone would reduce GHG emissions by more than 1.5 billion metric tons of CO_2 from business-as-usual annually by 2010. The plan also includes measures to: increase the use of renewable and nuclear energy; recover and use methane from coal beds, coal mines and landfills;

increase the development and use of bio-energy; utilize clean coal technologies; improve agricultural practices; and plant forests.

South Africa has analyzed a number of long-term mitigation scenarios. It has announced its intent to peak its emissions no later than 2025 and expects to have a final domestic climate policy adopted by the end of 2010. South Africa also continues to implement sustainable development policies and measures that will reduce GHG emissions. These policies and measures include moving from traditional coal-fired electricity production to renewables, nuclear power and clean coal technologies, improving energy efficiency and improving the efficiency of the transportation system.

Brazil has released a climate plan that emphasizes energy efficiency and reducing emissions from deforestation, including a goal to reduce the average deforestation rate by 70 percent over the period 2006-2017. It would lower CO_2 emissions by about 413 million metric tons CO_2 in 2010 (roughly one quarter of the emissions reduction expected in the Lieberman-Warner bill by 2015) and by a total of 4.8 billion metric tons CO_2 over the 12-year life of the program.

South Korea intends to announce a long-term, economy-wide target for emissions reductions later this year.

What will the global climate deal look like and how will international negotiations unfold?

In Copenhagen, developed or Annex I countries, including the U.S., are expected to agree to national, quantified GHG emission reduction targets. The European Union has already committed to reduce emissions 20 percent below 1990 levels in 2020 on its own, and increase its target to 30 percent below 1990 levels if other countries join.

U.S. engagement and commitment is critical for reaching a deal in Copenhagen. One only needs to look at the impact of the United States' recent decision to reverse its position and support the development of a new international agreement to reduce mercury emissions[1] to understand the implications of U.S. engagement. Almost immediately after the U.S. decided to support the development of a new agreement, China and then India supported the process as well.

Both developed and developing countries will judge U.S. leadership and commitment at Copenhagen on two criteria. First, has the U.S. committed to significant emission reduction targets? The stronger the proposed U.S. target, the greater the likelihood of stronger developing country actions. Although it would be ideal if the U.S. could pass domestic legislation setting out its emissions reduction targets before Copenhagen, in my view that is not necessary to reach a deal in Copenhagen. What is needed is sufficient action in both the House and Senate to give our negotiators a good sense of where our national cap is likely to be set.

The debate on acid rain legislation and the original cap and trade program for sulfur dioxide in 1990 may offer some useful historical insight. Senate and House legislative proposals quickly converged on the President's proposed cap level in 1989, the first year of President George H.W. Bush's term. The real battle raged over distribution of the allowance value among companies and regions which required another full year of debate, a pattern that could be repeated in the carbon debate this year. But the bottom line is that the critical piece for the international process is a consistent signal from the Congress on the cap level for U.S. negotiators to bring to the rest of the world to help reach the needed agreement in Copenhagen.

Second, has the U.S. committed to providing meaningful financing, technology and capacity building assistance to developing countries as it agreed to consider in the Bali Action Plan? As discussed earlier, each developing country is expected to take NAMAs — some unilateral and others conditioned on assistance. The specific details of what actions they will take in exchange for assistance will be addressed after the agreement in Copenhagen. The agreement in Copenhagen will establish the framework and policy architecture for developing country actions.

Some observers incorrectly assume that any financing agreement in the Bali Action Plan must mean large unrestricted amounts of funding. However, the behind the scenes negotiations are more likely to focus on specific and tailored financial mechanisms like support to "write down" the cost of advanced but not yet commercial technologies like carbon capture and storage, and financing for special purpose entities that can help overcome resistance from banks in developing countries to make financing available for energy efficiency. The European Commission has proposed the creation of a "facilitative mechanism" by which developing country proposals for action and specific requests for assistance can be evaluated based on objective criteria. The idea of "block grants" and the like are not under serious consideration.

Two additional issues will play an important role in the negotiations of the post-2012 framework: Reduced Emissions from Deforestation and Degradation (REDD) and adaptation. These issues will be important because they touch a much larger group of developing countries compared to industrial mitigation, where six to ten of the largest emitters will likely dominate the field. In addition, emissions related to deforestation and degradation are responsible for approximately 20 percent of global GHG emissions. Addressing these problems in a constructive way in the post-2012 climate agreement will be critical to solving the climate problem and will provide an important avenue for many developing countries to participate in the international effort to fight global warming. Likewise, adaptation affects virtually all countries, but has a particularly large impact on the poorest developing countries since they face the largest adverse impacts and have the least capacity to adapt to climate change. At Poznan, negotiators made progress on both REDD and adaptation. Reaching early agreement on the approach to these two issues early in 2009 will be an important building block for the larger Copenhagen agreement.

The level and extent of actions to reduce GHGs by developing nations in the post-2012 agreement is not only a critical question for the international debate, but also central to the outcome of the domestic debate here in the U.S. There is a great deal of concern in the U.S. with ensuring that U.S. companies are not placed at a competitive disadvantage if the U.S. takes action and other countries do not. The European Union has similar concerns.

There are two approaches under consideration in the U.S. and in Europe to address competitiveness. One would require border allowance purchase requirements (essentially a border tax) on imports from countries or sectors that have not taken comparable action to regulate GHG emissions. The other involves giving free allocations of allowances to those domestic companies in sectors facing considerable international competition, such as iron and steel, cement, pulp and paper and aluminum. The most interesting of these approaches in the U.S. is a proposal, H.R. 7146, that Congressmen Inslee and Doyle introduced in the 110th Congress, which would compensate domestic industries for the direct and indirect (energy) cost increases from carbon regulation they face until developing countries require the same industries in their countries to take comparable action to reduce GHGs. One benefit of this approach is the positive incentive it sends to cleaner companies within the U.S.

I believe both of these approaches could level the carbon cost playing field and can be viewed as complimentary, though under WTO rules we need to insure that the use of these measures either in combination or sequentially does not overcompensate U.S. industry and constitute protectionism. It is probably best to think about using the output based free allocation as the first line of defense with the border adjustment as a backstop. This is how the European

Union is approaching these two strategies, as the border tax adjustment is seen as provocative and could potentially trigger larger trade disputes.

I believe that it could make sense to operate such a program on a sector basis. The program would begin with output based rebates covering both the direct and indirect energy price increases facing our domestic industries in internationally competitive sectors. The portion of the rebate associated with direct costs of carbon would be phased out on a sector basis as a majority of the major emitting countries in that sector took comparable action. At that point, the border allowance adjustment would phase in for those other countries whose sectors had not taken similar action. The indirect energy cost portion of the output based rebate would continue until developing countries take action to reduce GHGs from their electricity sector or to establish a carbon price across the economy.

Although both of these strategies individually or in tandem could effectively level the carbon playing field, they will not create incentives for developing countries to reduce their domestic emissions or to cooperate in the negotiations. For example, according to a recent World Resources Institute and Peterson Institute Study[2], China exports approximately 8 percent of its steel production and exports only 1 percent to the U.S. It is unrealistic to expect that a border adjustment on 1 percent of Chinese steel would be a sufficient motivator for China to regulate the emissions from its domestic steel industry.

In my view, U.S. domestic legislation must also include provisions to encourage developing countries to take additional actions. Initially, this will involve creating incentives for them to reduce the rate of growth of their emissions to lay the foundation for absolute emissions reductions in the future.

One framework for providing incentives that has been garnering support internationally would rely on establishing the NAMAs discussed earlier in my testimony in key internationally competitive industrial sectors. This concept is included in the Bali Action Plan as "cooperative sectoral approaches and sector-specific actions" which are part of the actions suggested for mitigation of climate change. Under such sectoral approaches, developing countries would be asked to take a new commitment to reduce GHG emissions in a given industry sector beyond any recent unilateral actions they may have already adopted. They could receive up-front financial and/or technology incentives from developed countries in return. Mexico's announcement in Poznan of sectoral targets for key industrial sectors coupled with a 4-sector cap and trade program is the first concrete example of how such an effort might proceed.

Technology and finance assistance could be provided to developing countries by developed countries for a number of purposes. For example, assistance could be dedicated to build first-of-a-kind advanced technologies, such as carbon capture and storage, which are not yet cost effective, to accelerate technology deployment by bringing down the cost of advanced technologies, and as an incentive for participating developing countries to establish more aggressive "performance goals." This approach also creates opportunities for leading U.S. companies to gain access to growing new markets (creating jobs at home) and moves toward leveling the playing field for carbon in internationally competitive sectors.

In conclusion, with timing running short to avoid the worst effects of a warming planet, reaching an agreement on a post-2012 global framework for reducing emissions is crucial. Never before has the opportunity for a true global accord, involving all nations, been so close. It is clear that developing countries are already taking significant actions and that for the first time they are willing to take additional actions as part of an international agreement. What is needed is strong U.S. leadership demonstrated by a significant commitment to reducing emissions and providing assistance to developing countries. One should not underestimate how firm U.S. action will induce strong developing country action. The U.S. holds the power to unleash a race to the top that could overcome years of international inertia and leave a legacy to future generations for which all of us can be proud.

End Notes

[1] "Final Omnibus Decision on Chemicals Management" (UNEP/GC/25/CW/L.4) adopted by Twenty-fifth session of the Governing Council/Global Ministerial Environment Forum.

[2] Source: Peterson Institute and World Resources Institute 2008. *Leveling the Carbon Playing Field*.

Chapter 8

STATEMENT OF LEE LANE, RESIDENT FELLOW AT THE AMERICAN ENTERPRISE INSTITUTE, BEFORE THE SELECT COMMITTEE ON ENERGY INDEPENDENCE AND GLOBAL WARMING, HEARING ON "PREPARING FOR COPENHAGEN: HOW DEVELOPING COUNTRIES ARE FIGHTING CLIMATE CHANGE"

Mr. Chairman, Mr. Sensenbrenner, other members of the Committee, thank you for the opportunity to appear before you today. I am Lee Lane, a Resident Fellow at the American Enterprise Institute. AEI is a non-partisan, non-profit organization conducting research and education on public policy issues. AEI does not adopt organizational positions on the issues that it studies, and the views that I express here are mine, not those of AEI.

Rising amounts of greenhouse gases (GHGs) in the atmosphere pose worrisome challenges. While many uncertainties persist, I believe that the potential risks from climate change could be large. At the same time, a thicket of intractable problems blocks quick or easy solutions. Progress on climate policy will require us to wrestle with these problems over many, many decades. My statement suggests some ways in which the US might make progress on this task. It makes three main points.

First, we need to acknowledge that a seeming global consensus on the need to halt the rise in GHG levels masks a distinct lack of consensus on willingness to pay the required costs. Thus, many nations, China, India, and Russia prominent among them, reject all demands that they shoulder this burden. That China is willing to announce some no-regrets climate policies is a good thing. It will be better still if it implements them – something that is far from certain.[1] But a grab bag of marginal policy innovations is not something that will alter the basic realities of the problem.

Second, the dismaying truth is that the US cannot create global consensus where none exists. Efforts to lead by the example of stringent GHG reductions will be self-defeating. They would, in effect, make winning future concessions from China and India even more difficult. Conversely, current proposals to use trade sanctions to bludgeon other nations into adopting controls are too weak to compel such costly action.[2] And such proposals pose significant risks to the global trade regime.

Despite Chinese and Indian demands, having developed countries simply pay the costs of developing country GHG controls is not a viable option. A recent MIT study estimated that carrying this principle to its logical conclusion would, for the US alone, entail *annual* income transfers of $200 billion by 2020 and of nearly $1 trillion by 2050.[3] This would heap a huge additional burden on a US economy that, during this same period, will already be struggling to make many other daunting structural adjustments.

Third, the US government's climate policy should explicitly recognize the substantial likelihood that the needed global consensus on GHG curbs will be long in coming. The world is very likely to miss today's ambitious targets for GHG stabilization. Rather than striving to do the impossible, US policy should place great stress on fostering relevant new technologies. One way of doing so would be place a modest, stable, and gradually rising price on GHG emissions. I include a statement from a recent conference at Stanford University that describes other essential steps toward achieving this goal.

The statement notes the need to expand R&D directed at adjusting to the climate change that is unavoidable. And it notes explicitly that part of that effort should be directed to exploring so-called geoengineering technologies. Adaptation, by whatever means, needs more attention.

Finally, I would like to conclude with a note of caution. The US, like most other nations, has an important stake in curbing global GHG emissions. But if our good intentions lead us to incur costs that exceed the benefits *to America*, those policies may not prove to be durable. A zigzag course on climate policy is likely to serve neither this nation nor the world.

THE MISSING CONSENSUS ON WILLINGNESS TO PAY FOR GHG CONTROL

Given twenty years of failure to achieve meaningful progress on global GHG controls, one should ask, what structural changes are required to produce a different outcome, and have those changes, in fact, occurred?

A Record of Futility

The year 2008 was the 20th anniversary of the first meeting of the International Panel on Climate Change (IPCC). The IPCC's goal is to solve the problem of warming. So far, it has failed. According to the US Energy Information Agency, global emissions of CO_2, the most important industrial greenhouse gas, currently exceed the 1988 level by over a third. The IPCC reports that through the last several decades the rise in atmospheric concentrations of CO_2 has sped up.

Many Europeans blame the United States for this failure. They are especially harsh in criticizing the Bush Administration's rejection of the Kyoto Protocol. Even among observers who regarded Kyoto as a bad deal for the United States, the brusque manner in which President Bush rejected the Protocol seems in retrospect to have been unwise.[4]

The fact remains, though, that for America, Kyoto's high abatement costs and imposition of large income transfers from the US to other countries would almost certainly have made the Protocol a net loss. A multi-model assessment found that, in 2010, Kyoto would have cost the US between 0.24 percent and 1.03 percent of GDP.[5] Yet Kyoto would have had virtually no impact on global climate.[6] So whatever cost Kyoto imposed would have brought almost no benefit. On balance, it is less surprising that the US ultimately rejected Kyoto than it is that a US president had signed it in the first place.

Moreover, even in Europe, emissions continue to climb.[7] Where greenhouse gas emissions have fallen, changes in economic structure may have played a bigger role than has climate policy. In light of the record, faith in Europe's oft-repeated promises of swift GHG reductions would seem to demand a certain degree of credulity.

This experience raises a question that is pertinent to this new phase of climate policy. Have the conditions that doomed Kyoto to failure actually changed?

The Necessity of Curbing GHG Growth from China and India

GHG control policies can only succeed if they are based on coordinated multi-national action. A metric tonne of CO_2 has the same effect on warming wherever it originates, and fifteen to twenty nations around the world are major sources of GHG discharges. Economic growth will steadily raise the number of major sources. It is only a slight exaggeration to say that each of these states possesses an effective veto over global GHG control efforts.

Certainly, both China and India have such veto power. It is physically impossible to halt the rise of GHG levels if Chinese and Indian GHG emission growth is not reined in. With unchecked GHG growth from China and India, holding atmospheric GHG levels below 550 ppm would require the industrialized countries to somehow begin to capture more CO_2 than they emitted – and they would have to do so within thirty-five years![8]

China and India have so far flatly refused to incur significant costs in the cause of GHG reduction. To the contrary, their efforts at GHG control have been confined to what are, in effect, "no-regrets" policies. Both countries reject all firm commitments to GHG reduction targets. Rather, they have both demanded that the developed world commit to paying them for any emission reductions that they might undertake.[9] To some degree, the Bali Action Plan has endorsed this principle.[10] Thus, the current stance of the Chinese and Indian governments would seem to pose a rather stark challenge to the credibility of the entire enterprise of global GHG curbs.

The Limited Net Benefits of GHG Controls

Part of the difficulty of forging an international GHG control accord can be traced to the high costs of curbing GHG emissions. If GHG cuts are deep and rapid, their costs are likely to exceed their benefits.[11] Studies have repeatedly confirmed that judgment.[12] To yield the greatest possible net benefits, GHG cuts should, therefore, start modestly and increase gradually over time. However, controls structured in this way avoid only part of the expected climate change.

In this regard, GHG controls contrast sharply with the control of ozone-depleting chemicals. With the latter, optimal controls yielded quite large net benefits. The much smaller net gain available from GHG control restricts the range of options for deal making.[13]

The International Politics of GHG Controls

No third party exists to enforce participation in GHG limitation agreements, to compel performance of agreed actions, or to set standards. International politics is a self-help world; there is no 911 to call.[14] One result is that, in dealing with global problems, nations often have an incentive to free ride on the efforts of others.

In the case of GHG controls, the fact that nations differ so much in the degree to which they have an economic interest in curtailing warming vastly complicates the quest for international cooperation. These differences give rise to two further problems.

First, because nations differ in their level of concern about warming, they also differ in their willingness to incur the costs of restraining GHG discharges. An oil exporting country with a cold climate such as Russia has a lot less to gain from effective GHG controls than would the Maldives or the Sub-Saharan nations. The latter countries are more vulnerable to climate change. While there are quite a few high-GHG emitting countries that are poor and middle income,[15] most poor countries are not major emitters.[16] The latter countries may be threatened by climate change, but, on their own, they can do virtually nothing about it.

Second, in practice, even high-GHG poor countries lack the economic resources either to pay for GHG abatement elsewhere or to compel richer countries to adopt controls. Also, many poorer nations prefer to protect themselves from warming through economic development rather than by seeking to restrict GHG discharges. For middle income countries like China and India, industrialization can boost the ability to adapt to climate change. Of course, it can relieve many more acute problems as well. For these countries, slowing growth in order to control GHG discharges may simply be a bad investment.[17]

Limits on US influence on Global GHG Control Arrangements

For as long as these considerations apply, neither the United States, nor any other country or power bloc, will be able to install an effective global GHG control regime. Many inventive GHG control boosters have propounded schemes for breaking this political impasse. Their efforts have failed. New ones are likely to meet the same fate until the causes of the impasse are removed, and removing those causes is likely to take time.

Unilateral Action as a Poor Means of Building International Consensus

Congress is considering bills which, if enacted, would subject the US economy to strict GHG controls. Strict US controls, their advocates claim, would cause other nations, China and India among them, to adopt similar measures. Neither the historical record nor a fair reading of these countries' economic self-interest supports these claims.

As to the record, the Chinese government now claims that it will take some steps that will have the effect of lowering GHG discharges. The measures by which it proposes to make these reductions are classic no-regrets policies. Based on its statements at Bali, it has also offered to make further GHG reductions, but it appears to be willing to do so only if other nations pay much of the cost and transfer to China a great deal of advanced technology. India, at Bali, demanded that developed countries pay it for mitigation (and adaptation) as a matter of right. India also wishes to deny developed countries the ability to attach conditions to this aid.[18] This record strongly indicates that China and India place much greater stress on avoiding the costs of GHG abatement (and on gaining access at below cost to foreign technology) than they do on achieving the benefits of lowering emissions.

Economic self-interest, moreover, provides China and India with strong motives for resisting GHG limits. For example, as suggested above, from the standpoint of these countries, rapid economic development may well be a better response to climate change than GHG controls would be. Then too, neither China nor India may feel that its government enjoys enough popular support to be able to afford the political costs of GHG controls. To drive domestic energy prices above world market levels would be a daring political gamble for governments that have often gone to great lengths to hold those prices *below* world levels.[19]

America's unilateral adoption of stringent GHG limits would do nothing to weaken China's or India's motives for resisting controls. To the contrary, it would strengthen those motives. If the US adopts controls, it has thrown away a bargaining chip that it might have used in a future negotiation. Worse, the more other countries adopt GHG limits, the greater the competitive gains that China and India will reap by keeping their economies unencumbered by such controls. Over time, energy-intensive industries will migrate to the nations that resist GHG limits. And the increased concentration of energy-intensive capital and jobs in these countries will bolster the political incentives for resisting controls.[20]

If avoiding abatement costs is, indeed, a strong motive for China and India, the US would have little reason to believe that it could enforce an agreement – even if it made one. The seemingly perpetual problems with WTO enforcement illustrate that point all too well. The 2008 report on China of the US Trade Representative, while noting areas of progress, offered the following observation:

> "... in some areas it appears that China has yet to fully implement important commitments, and in other areas significant questions have arisen regarding China's adherence to ongoing WTO obligations, including core WTO principles. Invariably, these problems can be traced to China's pursuit of industrial policies that rely on excessive, trade distorting government intervention intended to promote or protect China's domestic industries. This government intervention, still evident in many areas of China's economy, is a reflection of China's historic yet unfinished transition from a centrally planned economy to a free market economy governed by rule of law."[21]

Similarly, the US government struggles – without great success – to ensure the safety of imported Chinese toys and food. Yet China clearly regards both the WTO regime and the reputation of its goods for safety as vital to its economic future. If China's *dirigiste* tradition and its weak rule of law disrupt its compliance with these trade-related regimes, how well would it implement GHG limits that it had been adopted only grudgingly?

The answer is not hard to guess. Further, America's record in enforcing product safety on Chinese products does not build much confidence on that side of the issue either. Could the US government determine, say, if a cap-and-trade scheme had been enforced on state-owned enterprises in Qinghai? Would we be able to tell if the cap had been offset with concealed subsidies? Meanwhile, India seems very concerned to make sure that the developed countries have no real ability to curtail GHG-related transfers for nonperformance or for any other reason.[22]

Trade Sanctions Will Not Lead to Chinese or Indian GHG Controls

One response to the China-India problem has been to propose to allow the US government to clap limited trade sanctions on other countries that fail to cap their GHG discharges. Such sanctions, their proponents maintain, would protect America's most energy-intensive industries from import leakage. They also hope that sanctions would prod China and India to adopt their own controls.

As a means of coercing China, this strategy would face long odds. There are two reasons for doubting that the incentive would work. China provides a perfect example of the grounds for skepticism.

First, China, by adopting domestic GHG controls, would handicap the competitiveness of most of its tradable products. This step would discourage exports and encourage imports. China's only compensation for accepting this result would be to eliminate US trade sanctions on a part of its trade with the United States. Moreover, the trade that might benefit from this substitution is a very small part of the Chinese economy. Less than 1 percent of Chinese steel production is sold to America in a form that would make it liable to sanctions. For aluminum, the number is only 3 percent. It is 2 percent for paper and less than 1 percent for both basic chemicals and cement.[23]

Second, one country adopting trade sanctions, or a few countries doing so, will merely change the geographic pattern of trade flows. It will not have much impact on the total demand for Chinese energy-intensive goods. US sanctions on China would cause countries with low-carbon processes for producing steel, aluminum, or other energy-intensive intermediate goods to increase their exports to the US. These countries could increase their own imports from China to fill the gap left by their higher exports. The Chinese would be largely indifferent to the change in trade flows. This option does not pose a serious economic threat to China, and it would certainly not compel China to adopt GHG controls.[24]

While sanctions would have little effect on China, they might threaten other US interests. For example, the precedent that they would set could further weaken the already fragile global trade regime. The threat is especially real given the tendency of the American political process to expand and escalate the effects of legislation that creates opportunities for restricting imports. The history of anti-dumping laws illustrates the grounds for concern.[25]

It seems likely that the sanctions Congress has considered so far would be ineffectual in compelling other countries to adopt GHG controls. In principle, though, the US might devise and, if need be, deploy, stronger trade penalties. The problem with that approach is patent. Trade sanctions designed to punish China will also hurt Americans. They will harm consumers, retailers, freight carriers, and manufacturers that use imported parts. The greater the pain imposed abroad, the greater the likely costs at home.[26]

Will the Annex II Countries Pay to Reduce China's GHG Discharges?

Alternatively, the US could offer to pay for China's GHG reductions as well as its own. Without question, this strategy is the one preferred by China, India, and the G-77 countries – for obvious reasons. At Bali, China and the G-77 countries demanded that the developed nations pay half of a percent to one percent of GDP to cover the costs of curbing GHG emissions in the developing world. Their demands at the Conference also extended to the transfer of technology at concessionary prices. India's demands, based on the position that it adopted at the Bali Conference, seem, if anything, more open-ended.

From a US standpoint, however, this approach has little to recommend it. A free ride of this kind creates incentives that will retard the process of China and India accepting the need to shoulder a significant share of the costs of controlling their own GHG discharges. Indeed, the more willing the US appears to be to entertain proposals that it will pay to abate other nations' GHG discharges, the more it will change the motives of the would-be recipients of its generosity. Once the principle is strongly established, every poor country with a substantial GHG output will have a strong motive to display a studied indifference to GHG controls. (I fear that this is, in fact, precisely what is happening.)

A less yielding bargaining stance on the part of the developed world would, over time, discover a growing Chinese and Indian self-interest in GHG controls. India, after all, is among the most vulnerable countries in the world to climate change. China, too, may have much to gain from slowing the pace of warming.[27] To be sure, the likely effects of future change remain unpredictable, and judgments vary as to each nation's relative share of the risks. Still, the stronger the evidence becomes suggesting future harm, the stronger the motives of today's most resistant nations to undertake control measures.

Without stiffer bargaining by developed nations, it is difficult to see how a viable global control regime can emerge. Placing on the developed world the full costs of making deep GHG cuts is an implausible option – as the MIT study strongly hints.[28] That study examined the costs of reducing GHG emissions by 50 percent by 2050. It found that the policy of giving the developing countries a full free ride on abatement costs would, for the developed world, entail a loss in economic welfare of 2 percent in 2020 and 10 percent by 2050. US losses could be greater or smaller, depending on how the sacrifices were allocated among developed nations.[29]

The transfer payments alone create a significant burden. The MIT study describes the results of one scenario thusly:

> "This transfer might also be compared to market flows—purchases of allowances will become part of developed countries [sic] import bill. To maintain trade balance an increase in imports of permits would need to be balanced by a reduction of other imports or an increase in exports. For the US exports were about $120 to $155 billion per month in 2007-08. Assuming US exports maintained the same relation to (projected) GNP, they would rise to $175 to $225 billion per month in 2020, and $385 to $500 billion in 2050. The US purchase of allowances in those years (taking *Full comp-equal cost* as an example) would require a 10% to 13% increase in exports in 2020 to maintain trade balance, and 29% to 37% in 2050."[30]

These are very large policy-imposed challenges. And they would be imposed on top of the already daunting post-bubble economic challenges to save, produce, and export more. If successfully met, these climate challenges would only restore the US economy to the condition of the *status quo ante*.

These factors would not preclude an agreement in which the US incurred substantial costs in the service of GHG controls. They do suggest that a GHG control agreement in which the US incurs large costs to reduce Chinese GHG emissions may not be politically sustainable.

The Risks of Exaggerating Chinese Efforts on GHG Reduction

America has reason to applaud Chinese actions to reduce GHG discharges. This is true even if those actions are of modest significance. At the same time, China has valid reasons to pursue efficiency gains. It may well also have opportunities for some profitable use of renewables. These innovations might produce some GHG reductions. The carbon efficiency of the Chinese economy has been very low. Countless options exist, no doubt, for improving it. Perhaps, many of these could produce net benefits quite independently of any concerns about climate.[31] The Chinese, therefore, have an incentive to make these no-regrets changes while trumpeting them as a sign of good intentions on GHG control.

Some advocates, however, may be tempted to pretend to accept some mix of these Chinese policies and measures as constituting serious action on GHG controls. This willing suspension of disbelief might offer short-run advantages in the task of enacting domestic GHG controls. In the long-run, though, it is dangerous climate policy.

Deep US emission cuts will not be cheap. If they do not produce comparable responses from many fast-developing countries, their impact on climate will be small. For the reasons discussed above, they are unlikely to elicit this response. If they do not, there is a real chance of the US policies being perceived as failures. (Indeed, they most likely would be failures in the sense of producing costs in excess of their benefits.) The last thing that climate policy needs at this point is for America to lurch into hasty GHG reductions and then reverse course when the discovery dawns that other key players have no intention of copying its actions.

A REALISTIC APPROACH TO AMERICAN CLIMATE POLICY

To all appearances, then, a policy of idealistic leading by example on GHG reductions is a risky one. Yet few people at this point would propose to ignore the looming worries about climate change. There is no good alternative to attempting to devise an active climate policy but one that can be pursued within our great, but limited, national means.

The Need to Set a Realistic Pace for GHG Reductions

A GHG control policy is unlikely to succeed until China, India, and many other fast-developing countries become willing to shoulder a substantial share of the costs. The passage of time is likely to increase these countries' willingness to pay. In the past, as countries have become wealthier, they have been inclined to spend more on environmental quality, a tendency that economists have dubbed the environmental Kuznets curve. This tendency may well apply to China, India, and other similar states.

Also, over time, these nations may develop the legal infrastructure that would allow them to implement more cost-effective forms of GHG limits. Today, they almost certainly could not implement a GHG tax or cap-and-trade scheme. Or, if they did, it might not produce the desired results.[32] Using command-and-control rather than market-based policies can, though, greatly increase the social costs of reaching a given level of GHG reduction.

Finally, new technology is likely, over time, to lower the costs of GHG control. This expectation is one reason that economically optimal GHG control scenarios concentrate emission cuts in the relatively distant future. For this reason, too, China and India are likely to be more willing to make GHG cuts later, when they can get larger returns for their investments.

Take the Long View; New Technology is Central

The high costs of GHG abatement, then, are one key source of climate policy's intractability. Only new technology can offer a way out of this difficulty. US climate policy should act on this insight. It should make the search for relevant new technology its top long-term priority.

This policy choice entails two initiatives. One is a modest carbon tax or, perhaps, a so-called hybrid cap-and-trade system. This measure would encourage the private sector to commercialize new low-cost technologies for abating GHG emissions. Keeping the GHG price low would limit the potential for competitive harm – even if America's trading partners failed to imitate its policy.

Credibility is another value of a modest and gradually increasing carbon price. The adoption of goals based on very steep GHG cuts is likely also to create a different source of unnecessary costs. Legislation that, if fully implemented, would lead to very high future costs may be greeted with skepticism. Investors might speculate that, when the economic crunch arrives, future officeholders may relax the goals to avoid imposing high costs on influential constituents. In that case, businesses may well adopt a wait-and-see stance with regard to investments in new technology. The result would be to delay new technology's advance into the market rather than to speed it up.[33]

For a low price to have much effect on the course of technology, the controls must be as cost-effective as possible. In the case of GHG controls, a price-based system is more cost-effective than any other policy tool.[34] Thus, a modest carbon tax or, perhaps, a so-called hybrid cap-and-trade system, would be the best available policy tool for creating the desired price.

Achieving the needed large declines in abatement costs, though, will require more than a price on GHG output. Breakthroughs in basic science will be essential.[35] The private sector generally invests little in basic science, and

GHG limits will not change that fact.[36] Some form of government support for basic science will be necessary to ensure that this investment occurs.[37]

Government has often been tempted to short-change basic energy science in favor of large demonstration projects. It has also found it difficult to avoid wasteful stops and starts in funding.[38] These challenges will doubtless reappear as government wrestles with the technological aspects of climate change. In the best of circumstances, the innovation process is likely to be a slow one. Prudence would seem to caution against expectations of sudden success.

Give Priority to Adaptation

A substantial amount of climate change is inevitable. Past emissions have locked it into the climate system. Fortunately, much can be done to minimize the net social costs of this change. America is well-endowed with the resources required to make the needed adjustments.

Many of these adjustments can be left in the hands of the private sector and of state and local governments. They have strong incentives to undertake the needed changes. Today, though, they are hampered by lack of knowledge about how regional climates will change and on what time scale.[39] Generating and diffusing this kind of scientific knowledge should be a top priority of federal climate policy. Developing this knowledge will depend on a strong, non-ideological climate science program. New knowledge in this area would clearly boost the nation's long-term economic productivity.

The federal government may also need to reassess some of its own policies. For example, public subsidies to disaster insurance may promote too much private sector investment in high risk areas. Climate change could worsen the potential resource misallocations. This risk may merit further study. In other instances, federal policies may cause under-pricing of some water resources. Again, the prospect of climate change may well increase the value of the resources being misallocated. Issues like these occasion intense passions. Yet the scale of changes that these policy changes would entail is no greater than some of those that GHG controls would impose on energy consumers, and the policy changes might yield net benefits rather than net costs.

A family of technologies, known collectively as 'geoengineering', might provide an added tool for adaptation. The idea behind them is simple. When sunlight strikes the Earth's surface, greenhouse gases in the atmosphere trap some of the heat that is generated. A slight decrease in the amount of sunlight reaching the Earth's surface could, in principle, offset the warming. Scientists estimate that deflecting relatively small amounts of the total sunlight that strikes the Earth back into space would be enough to cancel out the warming effect of doubling the pre-industrial levels of greenhouse gases.[40]

Scattering this amount of sunlight may be fairly easy. Past volcanic eruptions have shown that injecting relatively small volumes of matter into the upper atmosphere can scatter enough sunlight back into space to cause discernable cooling. The 1991 eruption of Mt. Pinatubo reduced global mean temperature by about .5 degrees Celsius. This temperature reduction was apparent in just a few months and persisted for about three years.[41]

Some scientists propose, therefore, to use modern technology to create a carefully engineered analogue to this effect. Proposals to seriously study geoengineering are gaining adherents among climate policy experts. In late 2006, NASA and the Carnegie Institution jointly sponsored a high-level expert workshop on the subject. The workshop report observed that such distinguished scientists as Ralph Cicerone, Paul Crutzen, and Tom Wigley and prominent economists such as William Nordhaus and Thomas Schelling have long argued that the concept warranted further exploration.[42] Recently, an expert conference conducted at Stanford added the voices of several more distinguished economists to those who have called for further research on this option.[43]

Seek International Agreements Where Real Agreement Exists

For climate policy, domestic and international initiatives are both required, and they must be mutually supportive. Thus climate diplomacy should support the kind of initiatives that have just been described. In general, the United States will doubtless engage other nations on climate change. More than a few opportunities exist for useful agreements. It is just that a global pact on GHG caps with full trading of emission allowances is almost certainly not among them.

In the meantime, though, many options for climate-related agreements do remain. On GHG controls, for example, a "targets and timetables" approach within a "pledge and review" framework would seem to make monitoring compliance much easier. This approach would also make penalties of failing to perform agreed actions more credible.[44] Attempts to agree to limited GHG controls of this type might make progress where the more ambitious GHG control plans are doomed to fail. (The downside is that the reductions that they achieve are likely to be small compared to the demands of the most zealous proponents of steep cuts.)

Prospects for technology cooperation may also be good. In addition, the US may wish to coordinate with other industrialized nations in order to help to boost the adaptive capacity of poorer states. It is realistic to pursue these opportunities, and doing so may yield economic, humanitarian, and security benefits.

CONCLUSION

In conclusion, America needs an activist climate policy, but it also needs a realistic one. Climate change is a serious concern, but it is not our only one, nor even the most pressing. Our responses need to account for the limitations on our resources and our abilities to affect the preferences of other societies.

Realism demands a willingness to engage in hard bargaining, and bargaining, as always, requires an ability to look beyond what others say in order to measure their deeds and to assess their interests. It also requires patience. These qualities are important. If we neglect them, the American people will pay more than they need to and get less climate protection than they could have.

APPENDIX A

A Statement on the Appropriate Role for Research and Development in Climate Policy[*]

Kenneth J. Arrow, Linda Cohen, Paul A. David, Robert W. Hahn, Charles Kolstad, Lee Lane, W. David Montgomery, Richard R. Nelson, Roger Noll and Anne E. Smith

---

[*] The views expressed here represent those of the authors and do not necessarily represent the views of the institutions with which they are affiliated. This research was supported by the California Foundation for Commerce and Education.

The Reg-Markets Center focuses on understanding and improving regulation, market performance, and government policy. The Center provides analyses of key issues aimed at improving decisions in the public, private and not-for-profit sectors. It builds on the success of the AEI-Brookings Joint Center. The views expressed in this publication are those of the authors.

Executive Summary

A group of economists and scientists met at Stanford University on October 18, 2008 to discuss the role of research and development in developing effective policies for addressing the adverse potential consequences of climate change. We believe that climate change is a serious issue that governments need to address. We also believe that it is vitally important that research and development be made a central part of governments' strategies for responding to this challenge.

A group of economists and scientists met at Stanford University on October 18, 2008 to discuss the role of research and development (R&D) in developing effective policies for addressing the adverse potential consequences of climate change. We believe that climate change is a serious issue that governments need to address. We also believe that research and development needs to be a central part of governments' strategies for responding to this challenge. Solutions to manage long-term risks will require the development and global deployment of a range of technologies for energy supply and end-use, land-use, agriculture and adaptation that are not currently commercial. A key potential benefit of focused scientific and technological research and development investment is that it could dramatically reduce the cost of restricting greenhouse gas emissions by encouraging the development of more affordable, better performing technologies.

Broadly speaking, economists identify three ways in which government can constructively address climate change. One is by pricing the damages caused by emissions leading to climate change. Doing so would induce individuals and firms to take better account of these damages in their everyday decisions. A second is through government research and development policy aimed at stimulating the search for new knowledge that could lead to breakthroughs in greenhouse gas reducing technology. A third is by taking and encouraging actions that would reduce the damage caused by greenhouse gas emissions. Here too, R&D can contribute by addressing technological means of damage-mitigation, including adaptation and geo-engineering. However, governments' support for technology R&D should cease at the development stage or in select cases the pilot demonstration phase. Risks and rewards from commercial deployment should be left for markets to determine, including, of course, whatever additional price signals arise from market-based mitigation policies.

The group agreed to the following set of principles as a guide to the design of an effective research and development policy for addressing climate change.

The Need for R&D Policy in Addition to Cap and Trade, Tax, Standards or Other Policies to Reduce Emissions

- An effective strategy to deal with greenhouse gas emissions requires that individuals and firms have incentives to take action to reduce their emissions. However, adequate control of greenhouse gas emissions almost certainly will require policies beyond pricing greenhouse gas emissions (or regulatory policies with the same end) and needs to include significant levels of direct and indirect support for basic and applied R&D.
- The payoff from effective R&D to reduce the cost of lowering greenhouse gas emissions could be very high.

The Need for Stable, Long-term Commitment to R&D Support

- Policy commitments must be stable over long periods of time. Climate change is a long-run problem and will not be solved by transitory programs aiming at harvesting availableshort-run improvements in energy efficiency or of low-carbon energy. A much morestable commitment to funding and incentives for R&D is required to do better than thelimited results of energy R&D efforts in the 1970s and 80s.
- Businesses and consumers must have credible and appropriate incentives for innovation if they are to develop new technologies that will be needed to mitigate and adapt toclimate change. Challenges include providing adequate funding for basic and fundamental

research, encouraging risk-taking, and promoting open access to information.
- Stable long-term commitments to R&D funding and incentives will change the direction of R&D.
- Among the steps governments need to consider in addressing such a long-term challenge are not just those that apply existing capabilities to climate-related research today, but also those that build the fundamental capacity to perform research in the future. This could include steps to promote training of scientists and engineers, rejuvenate laboratory capabilities in universities, and to establish programs to disseminate research information for example through internships, post-doctoral fellowships and exchange programs both nationally and internationally.

Design of R&D Programs

- Government R&D policy should encourage more risk-taking and tolerate failures that could provide valuable information. This can be accomplished by adopting parallel project funding and management strategies and by shifting the mix of R&D investment towards more "exploratory" R&D that is characterized by greater uncertainty in the distribution of project payoffs.
- The single greatest impediment to an R&D program that is directed at achieving a commercial objective is that it will be distorted to deliver subsidies to favored firms, industries, and other organized interests. The best institutional protections for minimizing these distortions are multi-year appropriations, agency independence in making grants, use of peer review with clear criteria for project selection, and payments based on progress and outputs rather than cost recovery.
- Technological progress requires both R&D and learning, so that R&D programs should not be planned in isolation from practical application. R&D can be required to make even a relatively well-developed technology suitable for particular applications, and attempts to make practical use of a technology can reveal points where additional R&D would be most productive.
- Climate change cannot be halted without technologies that are applicable to developing countries. Developing these technologies and

facilitating their adoption will likely requireengagement of R&D networks in developing countries.
- Research on how societies can better adapt to the effects of climate change and research on geoengineering as a measure to moderate temperature increases and climate impactsshould be included in a complete research portfolio.

The Limited Role of Technology Standards and Subsidies

- Mandatesand subsidies aimed at supporting the deployment of relatively mature technologies are unlikely to be cost-effective tools for eliciting the major reductions of greenhouse gas emissions that now appear to be called for. In some cases, performancestandards have proven effective in promoting engineering improvements and the wider adoption of existing techniques. Since the process of technology innovation anddiffusion can require an extended period of time, performance standards with shortercompliance periods cannot be expected to stimulate major breakthroughs.
- Technology-forcing performance standards have had a mixed record in inducing innovation. Regulators can find it difficult to obtain information about the status of technologies that is accurate enough to allow them to set standards that both can be achieved and will induce real innovation. Such standards may be effective when the pathto a technological solution is reasonably clear, but are less likely to be effective instimulating cost-effective and broad-based breakthrough technologies. This is especiallyrelevant in dealing with a multi-decadal issue such as climate change, where the challenge is to evolve standards with time in light of new knowledge and experience.

Kenneth J. Arrow
Stanford University

Linda Cohen
University of California, Irvine

Paul A. David
Stanford University

Robert W. Hahn
American Enterprise Institute

Charles Kolstad
University of California, Santa Barbara

Lee Lane
American Enterprise Institute

W. David Montgomery
Charles River Associates

Richard R. Nelson
Columbia University

Roger Noll
Stanford University

Anne E. Smith
Charles River Associates

End Notes

[1] Finamore, Barbara and Alex Wang. "Sticking to a Truly "Green" Stimulus." *ChinaDialogue*, 20 January, 2009.

[2] Houser, Trevor, Rob Bradley, Britt Childs, Jacob Werksman, and Robert Heilmayr. *Leveling the Carbon Playing Field: International Competition and US Climate Policy Design*. Washington, DC: Peterson Institute for International Economics, World Resources Institute, 2008; 76.

[3] Jacoby, Henry D., Mustafa H. Babiker, Sergey Paltsev, and John M. Reilly (2008). "Sharing the Burden of GHG Reductions." *MIT Joint Program on the Science and Policy of Global Change*, Report No. 167.

[4] Lane, Lee. *Strategic Options for Bush Administration Climate Policy*. Washington, DC: AEI Press, 2006.

[5] Hourcade, Jean-Charles and Priyadarshi Shukla. "Global, Regional, and National Costs and Ancillary Benefits of Mitigation" in *Climate Change 2001: Mitigation*. Contribution of Working Group III to the Third Assessment Report of the Intergovernmental Panel on Climate Change, B. Metz, O. Davidson, R. Swart and J. Pan (eds). New York: Cambridge University Press, 2001; 537, Table 8-8.

[6] Nordhaus, William D. and Joseph Boyer. *Warming the World: Economic Models of Global Warming*. Cambridge: The MIT Press, 2000; 152.

[7] Abboud, Leila. "EU Greenhouse-Gas Emissions Rose 1.1% Last Year." *Wall Street Journal*, 3 April, 2008; A8.

[8] Yang, Zili and Henry D. Jacoby. "Necessary Conditions for Stabilization Agreements." *MIT Joint Program on the Science and Policy of Global Change*, October 1997; 4.

[9] United Nations Framework Convention on Climate Change (2008). "Paper No. 3: Philippines on Behalf of the Group of 77 and China – Financial Mechanism for Meeting Financial Commitments under the Convention" in "Ideas and Proposals on the Elements Contained in Paragraph 1 of the Bali Action Plan;" 35-37; UNFCCC (2008). "Paper No. 1: Antigua and Barbuda on Behalf of the Group of 77 and China – A Technology Mechanism under the UNFCCC" in "Ideas and Proposals on the Elements Contained in Paragraph 1 of the Bali Action Plan;" 6-9; UNFCCC (2009). "China's Views on the Fulfillment of the Bali Action Plan and the Components of the Agreed Outcome to be Adopted by the Conference of the Parties at its 15th Session"; UNFCCC (2008). "Supplemental Submission by India: Why Financial Contributions to the Financial Mechanism of the UNFCCC Cannot be Under the Paradigm of 'Aid'".

[10] Jacoby, Henry D., Mustafa H. Babiker, Sergey Paltsev, and John M. Reilly (2008). "Sharing the Burden of GHG Reductions." *MIT Joint Program on the Science and Policy of Global Change*, Report No. 167.

[11] Kelly, David L. and Charles D. Kolstad. "Integrated Assessment Models for Climate Change Control" in *International Yearbook of Environmental and Resource Economics 1999/2000: A Survey of Current Issues*, H. Folmer and T. Tietenberg (eds). Cheltenham: Edward Elgar, 1999; 19.

[12] Nordhaus, William D. *A Question of Balance: Weighing the Options on Global Warming Policies*. New Haven: Yale University Press, 2008.

[13] Barrett, Scott. *Environment & Statecraft: The Strategy of Environmental Treaty-Making*. New York: Oxford University Press, 2003; 379.

[14] Mearsheimer, John J. *The Tragedy of Great Power Politics*. New York: W.W. Norton & Company, 2001; 32.

[15] Weisbach, David A. (2009). "Responsibility for Climate Change, by the Numbers." University of Chicago Law & Economics, Olin Working Paper No. 448; 12, Table 2.

[16] Weisbach, David A. (2009). "Responsibility for Climate Change, by the Numbers." University of Chicago Law & Economics, Olin Working Paper No. 448; 18-19.

[17] Schelling, Thomas C. (2005). "What Makes Greenhouse Sense?" *Indiana Law Review* 38: 593.

[18] United Nations Framework Convention on Climate Change (2008). "Paper No. 3: Philippines on Behalf of the Group of 77 and China – Financial Mechanism for Meeting Financial Commitments under the Convention" in "Ideas and Proposals on the Elements Contained in Paragraph 1 of the Bali Action Plan;" 35-37; UNFCCC (2008). "Paper No. 1: Antigua and Barbuda on Behalf of the Group of 77 and China – A Technology Mechanism under the UNFCCC" in "Ideas and Proposals on the Elements Contained in Paragraph 1 of the Bali Action Plan;" 6-9; UNFCCC (2009). "China's Views on the Fulfillment of the Bali Action Plan and the Components of the Agreed Outcome to be Adopted by the Conference of the Parties at its 15th Session"; UNFCCC (2008). "Supplemental Submission by India: Why Financial Contributions to the Financial Mechanism of the UNFCCC Cannot be Under the Paradigm of 'Aid'".

[19] Lane, Lee and W. David Montgomery (2008). "Political Institutions and Greenhouse Gas Controls." *AEI Center for Regulatory and Market Studies*, Related Publication 08-09.

[20] Jacoby, Henry D., Ronald G. Prinn, and Richard Schmalensee (1998). "Kyoto's Unfinished Business." *Foreign Affairs* 77(4): 54-66.

[21] United States Trade Representative. "2008 Report to Congress on China's WTO Compliance." December 2008.

[22] United Nations Framework Convention on Climate Change (2008). "Supplemental Submission by India: Why Financial Contributions to the Financial Mechanism of the UNFCCC Cannot be Under the Paradigm of 'Aid'".

[23] Houser, Trevor, Rob Bradley, Britt Childs, Jacob Werksman, and Robert Heilmayr. *Leveling the Carbon Playing Field: International Competition and US Climate Policy Design*.

Washington, DC: Peterson Institute for International Economics, World Resources Institute, 2008; 76.
[24] Ibid.
[25] Ibid; 41.
[26] Barrett, Scott. *Environment & Statecraft: The Strategy of Environmental Treaty-Making*. New York: Oxford University Press, 2003; 327.
[27] Anthoff, David and Richard S. J. Tol (2007). "On International Equity Weights and National Decision Making on Climate Change." Working Paper FNU-127.
[28] Jacoby, Henry D., Mustafa H. Babiker, Sergey Paltsev, and John M. Reilly (2008). "Sharing the Burden of GHG Reductions." *MIT Joint Program on the Science and Policy of Global Change*, Report No. 167: 26.
[29] Ibid.
[30] Ibid; 21.
[31] Montgomery, W. David and Sugandha D. Tuladhar (2006). "Making Economic Freedom Central to the Asia-Pacific Partnership." CRA, International.
[32] Lane, Lee and W. David Montgomery (2008). "Political Institutions and Greenhouse Gas Controls." *AEI Center for Regulatory and Market Studies*, Related Publication 08-09.
[33] Montgomery, W. David and Anne E. Smith. "Price, Quantity, and Technology Strategies for Climate Change Policy" in *Human Induced Climate Change*, M. Schlesinger, H. Kheshgi, J. Smith, F. de la Chesnaye, J. Reilly, T. Wilson, and C. Kolstad (eds). New York: Cambridge University Press, 2005.
[34] Hubbard, R. Glenn and Joseph E. Stiglitz. "Letter to Senators John McCain and Joseph Lieberman." 12 June, 2003.
[35] Chu, Steven, quoted in: Broder, John M. and Matthew L. Wald. "Big Science Role is Seen in Global Warming Cure." *New York Times*, 12 February, 2009; A24.
[36] Montgomery, W. David and Anne E. Smith. "Price, Quantity, and Technology Strategies for Climate Change Policy" in *Human Induced Climate Change*, M. Schlesinger, H. Kheshgi, J. Smith, F. de la Chesnaye, J. Reilly, T. Wilson, and C. Kolstad (eds). New York: Cambridge University Press, 2005.
[37] Arrow, Kenneth J., Linda R. Cohen, Paul A. David, Robert W. Hahn, Charles D. Kolstad, Lee L. Lane, W. David Montgomery, Richard R. Nelson, Roger G. Noll, Anne E. Smith (2008). "A Statement on the Appropriate Role for Research and Development in Climate Policy." *AEI Center for Regulatory and Market Studies*, Working Paper 08-12.
[38] Ibid.
[39] Repetto, Robert (ed). *Punctuated Equilibrium and the Dynamics of US Environmental Policy*. New Haven: Yale University Press, 2006.
[40] Lane, Lee L., Ken Caldeira, Robert Chatfield, and Stephanie Langhoff. "Workshop Report on Managing Solar Radiation." NASA Ames Research Center, Carnegie Institute of Washington Department of Global Ecology: NASA/CP-2007-214558, 18-19 November, 2006. Report published by NASA in 2007.
[41] Ibid.
[42] Ibid.
[43] Arrow, Kenneth J., Linda R. Cohen, Paul A. David, Robert W. Hahn, Charles D. Kolstad, Lee L. Lane, W. David Montgomery, Richard R. Nelson, Roger G. Noll, Anne E. Smith (2008). "A Statement on the Appropriate Role for Research and Development in Climate Policy." *AEI Center for Regulatory and Market Studies*, Working Paper 08-12.
[44] Schelling, Thomas C. (2005). "What Makes Greenhouse Sense?" *Indiana Law Review* 38: 581-593; Barrett, Scott. *Environment & Statecraft: The Strategy of Environmental Treaty-Making*. New York: Oxford University Press, 2003.

CHAPTER SOURCES

The following chapters have been previously published:

Chapter 1 – This is an edited, excerpted and augmented edition of a United States Congressional Research Service publication, Report Order Code R40936, dated December 30, 2009.

Chapter 2 – This is an edited, excerpted and augmented edition of a United States Congressional Research Service publication, Report Order Code R41049, dated January 26, 2010.

Chapter 3 – This is an edited, excerpted and augmented edition of a United States Congressional Research Service publication, Report Order Code R40001, dated January 7, 2010.

Chapter 4 – These remarks were delivered as Statement of Edward J. Markey, before the House of Representatives Select Committee on Energy Independence and Global Warming, given March 4, 2009.

Chapter 5 – These remarks were delivered as Statement of Carter Roberts, before the House of Representatives Select Committee on Energy Independence and Global Warming, given March 4, 2009.

Chapter 6 – These remarks were delivered as Statement of Barbara A. Finamore before the House of Representatives Select Committee on Energy Independence and Global Warming, given March 4, 2009.

Chapter 7 – These remarks were delivered as Statement of Ned Helme, before the House of Representatives Select Committee on Energy Independence and Global Warming, given March 4, 2009.

Chapter 8 – These remarks were delivered as Statement of Lee Lane, before the House of Representatives Select Committee on Energy Independence and Global Warming, given March 4, 2009.

INDEX

A

abatement, 8, 34, 69, 155, 157, 158, 159, 161, 164
accounting, 34, 84, 127, 139
accreditation, 38
acid, 6, 67, 81, 82, 89, 148
adaptation, 44, 77, 92, 93, 95, 96, 97, 99, 101, 103, 104, 106, 107, 108, 128, 142, 148, 158, 165, 168
adjustment, 12, 31, 149, 150
aerosols, 109
Africa, 117, 120, 147
agencies, 52
agricultural sector, 83
agriculture, 6, 7, 8, 20, 24, 25, 27, 47, 50, 118, 168
air emissions, 26
air pollutants, 53, 106
air quality, ix, 141
airports, 11
alternative energy, 31, 54
American Recovery and Reinvestment Act, 52, 119
ammonia, 6, 81
annual rate, 60, 133
appropriations, 53, 92, 99, 170
Asia, 42, 105, 117, 174
assessment, viii, 2, 20, 87, 88, 98, 111, 117, 134, 155
Austria, 73, 100
authorities, 23, 38
authors, 167, 168
automobiles, 15, 87, 135, 137

B

background, 84, 87, 90, 91
banks, 148
bargaining, 158, 161, 167
Beijing, ix, 132, 136, 138, 141
benchmarking, 77, 84
benchmarks, 9, 77, 78
bicarbonate, 6
biomass, 23, 54
blame, 155
Bolivia, 96, 99
Brazil, 21, 23, 25, 59, 93, 94, 96, 101, 104, 108, 112, 115, 121, 124, 125, 126, 127, 128, 142, 144, 145, 147
Britain, 59
building code, 33
Bulgaria, 7

C

Cabinet, 34
candidates, 58
capacity building, 32, 148
carbon dioxide, viii, 2, 3, 8, 14, 15, 65, 67, 68, 82, 86, 99, 116
carbon emissions, 12, 16, 36, 66, 82, 112, 121, 122, 135
Census, 61
centigrade, 114

CFI, 71
challenges, x, 4, 104, 114, 153, 162, 165
City, viii, ix, 91, 99, 141
Clean Air Act, 53, 55, 82, 89, 106
clean energy, 22, 52, 128
clean technology, 74
Climate Change Science Program, 144
closure, 29, 84
CO2, 3, 6, 7, 8, 12, 13, 14, 15, 16, 17, 18, 23, 25, 26, 30, 34, 35, 36, 38, 39, 40, 43, 44, 48, 54, 55, 58, 61, 66, 67, 68, 73, 81, 83, 86, 87, 89, 90, 99, 100, 102, 104, 105, 108, 116, 127, 132, 133, 134, 135, 136, 137, 140, 144, 146, 147, 155, 156
coal, 12, 20, 21, 22, 27, 29, 33, 47, 50, 62, 106, 120, 133, 135, 138, 139, 140, 146, 147
cogeneration, 17, 22
coke, 6, 67
combustion, 6, 67, 81
commercial bank, 137
commodity, 24, 70, 90
commodity markets, 70
community, 58, 68
compatibility, 92, 99
compensation, 160
competition, 9, 75, 149
competitive advantage, 115
competitiveness, viii, 10, 12, 15, 75, 78, 89, 92, 99, 111, 149, 160
competitors, 4, 10
compliance, 6, 9, 16, 36, 52, 58, 61, 66, 67, 68, 69, 73, 74, 75, 80, 84, 85, 92, 94, 102, 103, 109, 159, 166
composition, 31, 108
conductor, 43
conductors, 48
conference, viii, 10, 22, 37, 91, 100, 154, 166
connectivity, 35
consensus, 92, 96, 154
conservation, 29, 38, 48, 135, 136
consumption, 5, 14, 25, 38, 48, 123, 124, 134, 135, 136
consumption patterns, 123

control measures, 161
convention, 94
cooling, 8, 166
cost, vii, 1, 9, 12, 14, 21, 32, 34, 53, 58, 60, 67, 70, 75, 80, 81, 83, 84, 85, 93, 112, 137, 148, 149, 150, 151, 155, 158, 162, 163, 164, 168, 169, 170, 171
cost saving, 53
Council of the European Union, 74
covering, viii, 38, 65, 67, 83, 102, 127, 150
critical infrastructure, 68
Cuba, 96
culture, 104
currency, 58, 59, 60, 61, 84, 90
Cyprus, 7, 88
Czech Republic, 7

D

data collection, 69
database, 32
deaths, 104
decoupling, 137
deforestation, 21, 23, 24, 25, 44, 59, 77, 92, 96, 106, 108, 121, 126, 127, 128, 147, 149
degradation, 92, 96, 149
demonstrations, 32, 106
Denmark, 5, 12, 74, 95, 106
Department of Energy, 16, 58
destruction, 24
developed countries, 5, 46, 79, 97, 112, 142, 145, 150, 151, 154, 158, 159, 162
developed nations, 145, 146, 161
developing nations, 145, 149
development policy, 168, 169
diplomacy, 166
directives, 68, 79, 87
disaster, 165
discharges, 156, 157, 158, 159, 161, 162
discrimination, 74
distortions, 75, 84, 170
diversification, 23
domestic agenda, 25
dominance, 133
donors, 106

draft, 9, 79, 96
dynamics, 69

E

economic activity, 47, 69, 123
economic crisis, 50
economic development, 28, 93, 157, 158
economic downturn, 24, 134
economic efficiency, 77
economic growth, 8, 69, 119, 129
economic incentives, 47
economic integration, 26
economic performance, 75
economic resources, 157
economic welfare, 161
economy, ix, 15, 23, 28, 40, 46, 48, 50, 53, 55, 83, 95, 112, 114, 115, 116, 117, 119, 120, 121, 122, 123, 126, 132, 136, 137, 139, 147, 150, 154, 158, 159, 160, 162
Education, 167
EEA, 70
election, 19, 26, 38
electricity, 7, 8, 14, 16, 17, 19, 21, 22, 27, 29, 31, 35, 43, 45, 46, 47, 48, 54, 78, 79, 84, 106, 114, 124, 125, 126, 134, 135, 137, 138, 139, 146, 147, 150
eligibility criteria, 81
emission, 2, 3, 5, 7, 8, 10, 11, 12, 13, 15, 18, 19, 20, 21, 24, 26, 32, 37, 38, 40, 44, 46, 48, 50, 51, 52, 53, 54, 57, 58, 60, 67, 68, 72, 76, 80, 82, 83, 86, 87, 88, 89, 103, 114, 120, 145, 146, 147, 156, 163, 164, 166
emitters, 98, 99, 112, 118, 121, 149, 157
energy consumption, 5, 11, 29, 30, 34, 123, 133, 134
energy efficiency, 14, 17, 19, 22, 32, 33, 34, 35, 38, 44, 47, 48, 49, 50, 52, 53, 104, 112, 119, 121, 123, 132, 133, 134, 135, 136, 137, 139, 147, 148, 169
energy management system, 13
Energy Policy and Conservation Act, 55, 63
energy recovery, 53
energy supply, 50, 124, 126, 133, 137, 139, 168

enforcement, 24, 29, 30, 35, 66, 82, 127, 132, 135, 159
engineering, 7, 17, 168, 171
environmental impact, 86
environmental issues, 132
environmental protection, 139
Environmental Protection Act, 27
Environmental Protection Agency, 26, 52, 63
environmental quality, 163
environmental standards, 10
environmental technology, 40
EPA, 35, 38, 52, 53, 54, 55, 56, 63, 82, 83, 84, 89, 106
equipment, 36, 37, 38, 47, 53, 132, 135, 137
equity, 75, 77, 144
Estonia, 7
ethanol, 25, 26
EU, v, vii, 1, 2, 4, 5, 6, 7, 8, 9, 10, 11, 12, 13, 14, 15, 16, 17, 44, 47, 56, 57, 58, 65, 66, 67, 68, 69, 70, 71, 72, 73, 74, 75, 76, 77, 78, 79, 80, 81, 82, 83, 85, 86, 87, 88, 89, 90, 172
Eurasia, 63
European Commission, 4, 6, 9, 10, 13, 15, 66, 68, 87, 88, 89, 148
European Community, 5
European Court of Justice, 58
European Parliament, 6, 58, 74, 79, 81, 87, 88, 89
European policy, 13
European Union, vii, 1, 2, 4, 5, 7, 10, 11, 12, 14, 15, 17, 39, 65, 67, 73, 86, 87, 88, 89, 90, 104, 105, 133, 142, 144, 147, 149, 150
exaggeration, 156
exchange rate, 4, 58, 59, 60, 61, 84
exercise, viii, 65, 68, 106
expertise, 115
experts, 22, 46, 53, 60, 132, 133, 166
exploitation, 24
exploration, 24, 166
exports, 29, 31, 36, 150, 160, 162
exposure, 20, 75
extraction, 47

F

factories, 6, 38
family income, 48
federal law, 49
fertilizers, 36
financial resources, 118
financial support, 44, 97
Finland, 5, 12
fixed rate, 17
flexibility, 8, 79, 80, 85, 102, 103
forest management, 24, 139
forest resources, 24
formula, 76
France, 2, 10, 11, 12, 58, 70
free market economy, 159
fuel efficiency, 26, 30, 33, 36, 135
funding, 16, 19, 27, 34, 50, 52, 53, 98, 103, 119, 134, 135, 148, 165, 169, 170
fusion, 48

G

gasification, 33, 138
GDP per capita, 8, 76, 126
Germany, 2, 13, 14, 101
global trade, 154, 160
globalization, 116
governance, 93, 132
government intervention, 159
government policy, 168
Greece, 8
greed, 144
Green Revolution, 140
greenhouse gas emissions, ix, 25, 44, 67, 73, 76, 79, 81, 83, 87, 90, 92, 93, 120, 126, 129, 131, 133, 135, 139, 155, 168, 169, 171
greenhouse gases, viii, x, 6, 14, 67, 73, 83, 91, 100, 109, 117, 142, 153, 165
grids, 14, 43
gross domestic product, 114
grouping, 97
growth pressure, 54
growth rate, 69, 137

guidance, 33, 103
guidelines, 6, 13, 32, 76, 98, 103

H

harmonization, 26, 78, 81
host, 4, 99
household sector, 16
housing, 7, 8, 48
human activity, 117
human rights, 104
Hungary, 7
hybrid, 43, 136, 164
hydroelectric power, 32
hydrogen, 6

I

ideal, 148
Impact Assessment, 104
impacts, vii, 1, 2, 3, 75, 78, 92, 93, 95, 99, 104, 114, 115, 117, 118, 119, 120, 121, 128, 132, 149
imports, 10, 12, 31, 36, 107, 149, 160, 162
incidence, 132
inclusion, 66, 81, 86, 106
income tax, 12
income transfers, 154, 155
independence, 170
Independence, v, vi, 111, 113, 131, 141, 153
India, 33, 34, 35, 36, 60, 61, 68, 93, 96, 105, 107, 112, 114, 115, 118, 121, 122, 123, 124, 125, 147, 154, 156, 157, 158, 159, 161, 163, 164, 173
indigenous peoples, 99
Indonesia, 115
industrial processing, 27
industrial sectors, 25, 79, 83, 145, 146, 150
industrialization, 128, 157
industrialized countries, 33, 94, 95, 100, 102, 106, 156
insight, 66, 82, 148, 164
intellectual property, 18, 107
intellectual property rights, 107
interference, 93, 101, 142
intervention, 159

investors, 34
Ireland, 6, 58, 80
iron, 6, 7, 11, 17, 31, 35, 67, 133, 140, 149
isolation, 170
issues, ix, x, 10, 21, 53, 57, 66, 68, 82, 87, 89, 92, 93, 96, 99, 103, 108, 120, 132, 141, 148, 153, 168
Italy, 80

J

Japan, 36, 37, 38, 39, 40, 61, 105, 106, 107, 137
justification, 128

K

Kazakhstan, 109
Kenya, 105
Korea, 40, 43

L

labeling, 14, 36, 38, 49, 135
land tenure, 127
landfills, 53, 146
Latin America, 117
Latvia, 7, 77, 80
law enforcement, 24, 127
leadership, ix, 47, 113, 120, 126, 128, 129, 142, 147, 151
leakage, 9, 58, 76, 78, 79, 89, 160
learning, viii, 65, 68, 170
Least Developed Countries, 97, 104
legislation, vii, ix, 1, 5, 7, 24, 38, 51, 53, 55, 92, 99, 104, 112, 113, 128, 141, 148, 150, 160
legislative proposals, 148
lifetime, 16
light-emitting diodes, 43
liquidity, 69
Lithuania, 7, 80
local authorities, 55
local government, 29, 33, 38, 45, 135, 165
logging, 23, 127

M

machinery, 38
majority, 11, 54, 63, 104, 128, 150
management, 33, 35, 106, 127, 132, 135, 170
mandates, viii, 8, 91, 92, 95, 96, 99, 107, 108
manipulation, 66, 80, 86
manufacturing, 22, 27, 30, 39, 43, 44, 47, 60
market economics, 66, 85
market share, 4, 43
marketplace, 70, 84
masking, 114
matrix, 22, 23
membership, 67
mercury, 27, 83, 133, 147
metals, 45
Mexico, viii, ix, 44, 45, 91, 99, 108, 112, 120, 123, 124, 125, 141, 142, 144, 145, 146, 150
modernization, 14, 48

N

national policy, 84
natural gas, 12, 14, 17, 21, 46, 47, 48, 50, 66, 86
natural resources, 25
neglect, 167
negotiating, viii, 91, 95, 96, 99, 108
Netherlands, 102, 140
nitrogen, 109
nitrous oxide, 2, 3, 81, 86, 109
North America, 44, 62
Norway, 12, 24, 86
nuclear energy, 43, 50, 146
nuclear power, 17, 147

O

Obama Administration, 55
objective criteria, 148
oil, 12, 19, 27, 43, 45, 46, 47, 48, 49, 50, 66, 67, 86, 119, 136, 146, 157
oil sands, 27

opportunities, 18, 44, 151, 160, 162, 166, 167
organic chemicals, 6
Outer Continental Shelf Lands Act, 55
oversight, 66, 86
ozone, 100, 109, 156

P

Pacific, 105, 174
paradigm, 40, 62
paradigm shift, 40
parallel, 50, 105, 170
parity, 92
Parliament, 19, 33, 34, 61, 79, 89
pathways, 121, 128
penalties, 8, 11, 16, 30, 103, 109, 160, 166
performance, 9, 16, 35, 36, 39, 43, 49, 54, 55, 70, 78, 136, 151, 157, 168, 171
performance benchmarking, 43
permission, iv, 7
permit, 9, 35, 80, 82, 84
pesticide, 11
Philippines, 125, 173
plants, 6, 14, 22, 29, 32, 35, 43, 50, 61, 67, 69, 138
pleasure, ix, 131
Poland, 7, 70, 77, 80, 96, 107, 146
policy choice, 164
political opposition, 19
politics, 157
polluters, 12
pollution, 8, 27, 28, 29, 35, 61, 104
popular support, 158
portfolio, 25, 54, 171
Portugal, 8
poverty, 93, 114, 128
power plants, 29, 33, 134, 137, 138, 140
precipitation, 102, 118
President Clinton, 94
prevention, 44, 109
price changes, 66, 86
price floor, 85
price signals, 84, 169
private investment, 45
private sector investment, 165

probability, 69
procurement, 14
producers, 7, 10, 31, 44, 52, 54, 58, 84
production capacity, 31, 32, 125, 134, 140
production costs, 58
profit, ix, x, 141, 153, 168
project, viii, 15, 32, 46, 52, 65, 68, 69, 72, 78, 85, 106, 127, 134, 138, 170
proliferation, 55
prosperity, 115, 117, 119
protected areas, 23, 127
protectionism, 12, 58, 149
public awareness, 136
public education, 53
public health, 133
public opinion, 136
public policy, x, 153
public sector, 16, 17, 48
public service, 136
pulp, 6, 67, 149

Q

quotas, 14, 17, 35, 43

R

Radiation, 174
raw materials, 31, 32
reading, 89, 158
reality, 112, 128
recession, 37, 40, 70
recognition, 116, 119, 121, 132
recommendations, iv, 140
reforms, 30, 45, 127
regeneration, 35
regulatory framework, 54
renewable energy, 5, 8, 14, 17, 22, 23, 25, 35, 38, 40, 43, 44, 46, 48, 50, 52, 77, 104, 121, 124, 125, 126, 138
replacement, 22, 138
requirements, 2, 3, 10, 21, 30, 31, 36, 49, 52, 56, 66, 68, 69, 70, 73, 74, 82, 92, 99, 145, 149
reserves, 23, 66, 68, 84, 86, 119
resistance, 148

resolution, 83
resources, 19, 23, 30, 31, 33, 47, 49, 50, 139, 165, 167
respect, 69, 74, 77, 79, 97
restructuring, 31, 122, 134
revenue, 8, 12, 16, 19, 20, 47
rewards, 145, 169
rhetoric, 85, 116
rights, iv, 15, 127
risk-taking, 170
Russia, 46, 47, 48, 49, 50, 62, 63, 104, 106, 154, 157

S

sanctions, 10, 154, 159, 160
savings, 14, 16, 22, 31, 36, 134
scale system, 19
scaling, 125
scarcity, 88
scatter, 166
scientific knowledge, 165
seasonality, 102
self-interest, 158, 161
Senate, 19, 51, 92, 94, 101, 104, 105, 107, 126, 140, 148
sensitivity, 3, 57
signs, 59, 112
small businesses, 37
social costs, 164, 165
social development, 47
social infrastructure, 14
social movements, 99
South Africa, 96, 112, 115, 120, 147
South Korea, 40, 43, 105, 147
sovereignty, 93, 97
space, 96, 135, 166
Spain, 6, 10, 58, 80
speculation, 23, 86
speech, ix, 10, 48, 61, 112, 132
Spring, 85, 118
stabilization, 120, 154
stakeholders, 52, 68, 96
standardization, 77
state-owned enterprises, 159
statistics, 31, 61

statutes, 53
steel, 6, 7, 11, 15, 17, 31, 35, 36, 39, 40, 43, 67, 78, 133, 140, 146, 149, 150, 160
steel industry, 31, 150
stimulus, 119, 139
storage, 19, 27, 77, 138, 148, 151
strategy, 14, 18, 33, 47, 50, 55, 160, 161, 169
stressors, 104
structural adjustment, 154
structural changes, 155
subsidy, 32, 77, 135
subsistence, 118
substitution, 22, 160
Sudan, 96
sugarcane, 22, 25
sulfur, 2, 3, 27, 67, 82, 86, 109, 148
sulfur dioxide, 27, 67, 82, 148
Supreme Court, 55, 106
surplus, 31, 46, 47
survey, 68, 136
survival, 118
sustainable development, 23, 24, 145, 147
Sweden, 5, 10, 12
Switzerland, 12

T

tariff, 31, 49
tax breaks, 34
tax incentive, 43, 52, 53
taxation, 6, 24
technical assistance, 3, 52, 135
technology transfer, 107
temperature, 5, 97, 98, 100, 102, 103, 108, 114, 117, 118, 166, 171
tenure, 127
territory, 7, 23
textiles, 35, 45, 140
threats, 114
Title I, 63, 67, 82, 86
Title II, 63
Title IV, 67, 82, 86
total energy, 124
tracks, 92, 95, 105
trading partner, 99, 164

trading partners, 99, 164
transfer payments, 162
transparency, 49, 97
transport, 5, 6, 7, 8, 13, 16, 19, 22, 32, 35, 48, 50
transportation, 8, 25, 26, 27, 43, 54, 67, 83, 136, 142, 147
trends, 24, 49, 50, 69, 128
tropical storms, 118
Tuvalu, 96

U

U.S. economy, 94, 101, 104
UK, 15, 16, 17, 18, 59
UN, 59, 100, 106, 115, 116
unemployment rate, 58
uniform, 74, 77
United Kingdom, 2, 15
United Nations, 12, 26, 49, 62, 91, 94, 96, 100, 108, 144, 173
United States, 175
universities, 52, 170

V

vegetation, 46, 95, 109
vehicles, 7, 10, 12, 17, 29, 30, 32, 33, 36, 37, 39, 40, 45, 52, 54, 55, 56, 136

Venezuela, 96
venture capital, 34
veto, 75, 156
vision, 17, 59, 92, 95, 97, 106, 107
volatility, 80, 85, 119
vulnerability, 44, 118

W

waste, 5, 7, 8, 19, 25, 27, 47, 48
waste disposal, 5
waste treatment, 25
water heater, 53, 135, 138
water resources, 132, 165
water supplies, 118
water vapor, 108
Western Europe, 104
White House, 51, 52, 63
wholesale, 135
wildfire, 118
wind farm, 14
wind turbines, 138
World Bank, 24, 97, 106, 135, 140
World Trade Organization, 4
WTO, 4, 10, 57, 89, 149, 159, 173

Z

zero sum game, 128